Dynamic Kabbalah

Dynamic Kabbalah

YOUR COSMIC CONNECTION

Gary Hanson

To order additional copies of this book, contact:
Xlibris Corporation
1-888-795-4274
www.Xlibris.com
Orders@Xlibris.com
17749

Contents

Chapter 1

WHY STUDY DYNAMIC KABBALAH?

For thousands of years the teachings from the Kabbalah have been preserved and carefully taught to a chosen few. The Kabbalah was taught orally by father to son, or by a teacher to a carefully selected student. It was necessary to guard the knowledge contained in the Kabbalah with such secrecy, so as to preserve it intact, unaltered and unpolluted. A teacher of Kabbalah would often spend a lifetime searching for the right student.

A student was usually at least 40 years old and had to pass a variety of tests before he was considered ready and able to learn Kabbalah. Once selected, the student would spend the rest of his life pursuing nothing but the understanding of Kabbalah. After his teacher died, the student would become the teacher in search of another worthy student. This process continued from generation to generation for hundreds of years until some of the teachings of Kabbalah were finally written down on parchment.

Kabbalah is sometimes spelled Cabala, Cabalah, or Qabala, depending on who is using it. Kabbalah comes from the Hebrew word KBL, or QBL, which means, *"to receive."* This refers to the knowledge that was received by the earliest representatives of mankind explaining who we are, how we got here, and what our creative capabilities are. Kabbalah also reveals our purpose in life upon this planet. Learning to use our vast creative capabilities is what originally made it necessary to bestow the precious knowledge only upon the most trusted, and tested, of individuals. Ignorance and misuse of the creative principles of Kabbalah is what has produced the Civilization that we know today. Our civilization has been maintained from one generation to the next by strict adherence to long established rituals and traditions.

Worldwide, there are thousands of differing religions, philosophies, and socio-political ideas from which to choose a way of life. Most people just grow up and accept whatever they have been taught and look no further. To learn something new requires change. Most people would rather suffer their present life, than make the effort to change it. To change one's circumstances, rituals, and traditions requires changing the way one thinks. It requires replacing old thinking patterns and habits with some that are new. And for most people this is a difficult and unpleasant experience. However, there are some people who are internally driven to search for something better. It is from these few people that students of Kabbalah are chosen!

Change is necessary for spiritual growth. Let's be honest about it. No one wants to make serious changes in their lives because it would be admitting that their beliefs and ideas are wrong. It might even mean that their whole system of logic is wrong. Hey! It might also mean that some of their most cherished convictions are faulty! What's more, what would their friends and family say if they began to *change* the way they think and believe? It is much easier for most people to continue to live in crisis, conflict, and chaos, than to try to find its cause and make the correction. It is easier for some to live with mismanagement, misery, and misfortune in their lives than it is to move to a better moment. Understanding dynamic Kabbalah can provide greater meaning to the changes that they desire in their lives.

"Crisis is a challenge to change."—David Richo Ph.D.

When I was a child I was raised in a church that today would be called "a religious right-wing church." Even as a child I felt driven to read my Bible so as to discover and understand what today is called "universal truth." I began to realize that my church held doctrines that did not support universal truth. When I discussed this with my ministers, they counseled that if I did not understand the Doctrine, I should just accept it on faith. But I couldn't. I had to look deeper into the matter. For me, it was a crisis. I was being challenged to change! No one seemed to see what I saw. I felt as if I were all alone in the world. In due time I found someone who understood as I did and was able to teach me more. It is an axiom that when a student is ready for further advancement, a teacher will come. It is normally an "unusual co-incidence" that puts a prospective student in touch with a spiritual teacher. As my spiritual and philosophical beliefs and ideas began to evolve, friends and family became concerned because they feared the changes in my life. Eventually this fear caused separation from some of my friends and family. I began to feel as if I were living an exciting new life. I soon discovered new friends and associates with whom I could share this new life.

When people are searching for truth, they are looking for food for their own soul. When they follow the new path they have just learned, their soul strengthens and grows. Conversely, when they stop searching, and do not reconcile what they have presently learned with what they have always believed, their soul gradually weakens and eventually falls asleep. A soul that is struggling to stay awake continues to re-assert itself. You will sense this internal search and reconciliation within your own mind when you struggle with a decision of what you consider to be right or wrong. The Apostle Paul described this ongoing battle between the soul and the mind in Romans 7:15-19, **"I do not understand what I do. For what I want to do I do not do, but what I hate I do . . . . For I have the desire to do what is good, but I cannot carry it out. For what I do is not the good I want to do. No, the evil I do not want to do—this I keep on doing . . . "**

My soul continued the struggle with my mind and body. As I grew older, I studied many different religions and philosophies. I saw many similarities among them, but also many differences. Years later, when I was introduced to the teachings of the Kabbalah, I began to see the foundation of universal truth. Everything began to make sense to me, and long held questions began to be answered for me. I began, slowly, to understand why the Universe was created and why humanity came to dwell upon the earth. I began to understand my purpose in the universal scheme of things. I also began to fathom the meaning of morality and what was right and what was not. And I started to lose the chaos and confusion that my life once held. If you feel as I once did, I hope your questions will soon begin to be answered as well.

David Rico Ph.D., in his book *Unexpected Miracles*, wrote, *"Chaos is . . . a condition for something new to emerge."*

When Mount St. Helen's volcano exploded and blew a large portion of its top into the air, and continued to erupt for days, there was total chaos in that area. Thousands of acres of trees were burned and blown down, leaving behind a desert of ash. However, to the surprise of many environmentalists, various forms of life immediately began to move back into the volcanic areas, and vegetation began to return as well. You see, *nature* is structured so that new life will emerge out of chaos.

The world of *chemistry* is similarly structured. A booster rocket can be powered by solid fuel. Liquefied oxygen and hydrogen can be combined to produce a controlled explosion, or chaos, that gives the rocket its thrust. A byproduct of that explosion is the combination of the oxygen and hydrogen into water vapor, which is a new product.

David Rico may not even know what the Kabbalah is, never the less his book reveals an important principle of Kabbalah. The events of your life are a result of cause and effect. Everything that happens to you has a cause. Your reaction to what has happened determines what will happen next and is the effect. Your perception of what happens to you determines what reaction you take in any given situation. And that is the story of your life!

This is best illustrated by the movie, *Groundhog's Day* in which the character kept living the events of a single day over and over again. Eventually his perception of certain events began to change and he was able to change some of the events to an outcome better to his liking. As his reaction to the events changed, so did the outcome of what happened next. Finally he discovered that his basic character and attitude had been changed as well.

Once you understand that your life story is a series of causes and effects, and your interaction with the results of those effects, then you can develop a better method of improving your own personal life. When you first began your spiritual journey you were searching for knowledge, and for answers to certain questions. But eventually you may come face-to-face with the realization that to continue on further, difficult changes will be necessary in your life. This realization then becomes a crisis. Do you change your life and go on, or do you turn back to the old predicaments, the old habits and the old ways?

When you get to this level in your studies remember that crisis is a challenge to change. And when the events in your life become chaotic you will need to confidently believe that you now have the opportunity for something new in your life to emerge.

Most people realize that we are beginning to enter the Age of Aquarius. In the previous age of Pisces that we have been experiencing for the past 2000+ years, material life, and things pertaining to matter, have overshadowed the spiritual side of life. Even things spiritual have been understood from a materialistic viewpoint. But this is slowly beginning to change. The understanding that our material existence is based in a series of *energy systems* is beginning to take root. As the old paradigm fades away, a new paradigm of learning how to manage our internal energy systems is gaining strength. A wide variety of "alternative" health products, and healing techniques have already been developed to aid in the management of our internal energy systems. To complete the revolution to the new paradigm we will be required to change the way we think and the way we perceive our environment and our place in it. Dynamic Kabbalah will aid you in making that transition.

It has been said that there are 3 kinds of people on the earth today.

People who *wonder* what the hell happened!

People who *watch* the things that happen!

People who *make* things happen!

Which kind of person do you want to be?

Steven Covey, in his book *7 Habits of Highly Effective People* explained a situation that he and his wife wanted to change concerning their relationship with a third person. He wrote, **"We began to realize that if we wanted to change the situation we first had to change ourselves, and to change ourselves effectively, we first had to change our perceptions."**

If you take the study of the principles of Kabbalah seriously you will begin to see a change in your perceptions. Your perception of the Universe, and of nature, and of the source of life will begin to change. Your perception of who you are and your true purpose in this life will also begin to solidify. In time, you may even begin to learn the Wisdom of Solomon and start to apply it to your own life. This deepening inner knowledge will begin to bring a sense of power and control to your own life, and even to your own environment. And as your perceptions change, so will the crises or chaotic events in your life begin to change.

Principles of Kabbalah reveal both the source and the mechanics of the Universe. You will learn that the Universe does not stand-alone by itself. You will discover that the Universe is the manifestation of an even more powerful energy-information system that powers it, sustains it, and holds it together. You will also learn, if you continue these studies, that the human mind does not exist as an entity unto itself. The human mind is actually an extension of the powerful energy-information system I have just described.

What does this mean?

For one thing, it means that *YOU are NOT ALONE.*

Once we gain a more solid understanding of the energy system that surrounds us, we will better understand how to benefit from it.

Shakti Gawain in *Creative Visualization,* wrote, **"We are especially susceptible and receptive to the thought forms that we hold about each other. It is these thought forms and the underlying attitudes that they reflect, which form our relationships, and cause them to work, or not to work."**

Thought forms are a more complete collection of thoughts that attach themselves to a feeling about someone. Think about a close friend or family member and what do you see? You see the person's face and form, and a positive or negative feeling about them. Attached to the feelings about them are thoughts about past events, or future fears, hopes, and dreams. The whole package that come to mind when you hear the person's name or think about them is a thought form. Your present religious belief, or personal philosophy, is also a thought form. Your political views consist of a thought form. Even your views on marriage are based on a thought form. Thought forms are solidified by your perceptions, which have been built up over a period of time. If you continue your study of Dynamic Kabbalah, some of your currently held thought forms may be shaken. Simultaneously, your reaction to certain knowledge might be, "I didn't realize it until now, but I think I have always believed that!"

There are dozens of books about the Kabbalah in the bookstores, but most of them give you only a portion of the knowledge of the Kabbalah. But using principles of Kabbalah from the Bible we will reveal things that you will never find in other books of Kabbalah. I have also used quotes from several authors of books that have nothing to do with Kabbalah, just to point out that principles of Kabbalah can be found almost anywhere, and especially in some of the latest *scientific* discoveries. From this information you will have the opportunity to develop your own life with the dynamics of Kabbalah.

Dynamic Kabbalah is in perfect harmony with most of the scientific discoveries being made today. As telescopes peer out to the very edge of the known Universe their startling discoveries can be explained by the dynamics of Kabbalah. And as electron microscopes delve more deeply into the heart of the atom, scientists come ever closer to viewing the energy-information system that sustains that energetic little atom.

A ton of research is being done on a wide variety of fronts in the fields of DNA and other cellular research. This can best be summed up by Paul Pearsall Ph.D. in his book, *The Heart's Code*. **"By being open to the possibility that cells remember, and that their memories may be deeply registered as a template of our soul, we may be able to understand more about what we often call 'our basic temperment.'"**

Each individual cell in your body is more complex than the most sophisticated supercomputer ever built by human hands. Cloning is the science of taking an individual cell and creating an exact duplicate of the entire organism. Remember *Dolly* the sheep? By using the cell's coded DNA information, an entire sheep was created and birthed. The entire pattern for the size and appearance of the cloned sheep was there. The instructions for growing from an individual cell into a mature animal were also contained in the DNA. Even the temperament of the donor animal was transferred to the clone through genetic information. The cloned animal looked exactly like its DNA donor, and even its behavior and temperament were very closely similar.

Every piece of information about your body and your brain is stored in the genetic coding of the *DNA* in every individual cell of your entire body. Any randomly selected cell of your body could yield the same general information as any other cell of your body. That is hard to believe, isn't it? Yet, the dynamics of Kabbalah can help you to understand how this is possible.

Would you believe that the cells in your body can also remember your life experiences, your character, and your temperament? Dr. Pearsall's research with heart transplant recipients has discovered that the transplanted heart brings with it a residual memory of people and events from the donor's life. Even the donor's Characteristics and temperament are often felt by the recipient. It is scientific research such as these that bring the world's fund of knowledge ever closer to the dynamics of Kabbalah.

The basic foundation of all of the world's major religions has one theme in common. When you trace their beliefs and truths back to their original founding, you will find many of the principles of Kabbalah. In many of the surviving religions of today, the principles of Kabbalah have been forgotten amidst the rituals that have endured from each generation to the next. The rituals have often been enriched by local customs that have led to greater diversities through isolation and extended time. However, with patient research, some principles of Kabbalah can still be found within every major religion.

Christianity has often thought of itself as a "new religion" beginning about 2000 years ago. But the concepts of Christianity were not new. The principles of Christianity have existed for thousands of years. Even the idea of a Messiah or a Savior, has been known since the beginning of time. All of the ancient Mesopotamian religions have had their versions of a God who reincarnated as a man so as to become a Savior to provide mankind with eternal life. And where did all of these ideas come from? You guessed it! From the ancient principles of Kabbalah.

The principles of Kabbalah form the very foundation of the Bible, and especially the New Testament. In recent years the Bible has fallen out of favor with many people because of its close association with the Christian Church. The Bible has been used to create a wide variety of divergent Theologies, and many of those theologies have proven to be out of step with scientific discoveries. Some theologies have adapted to scientific knowledge and others have not. In order to adapt, some theologies have just discarded the Bible completely. What a shame that is!

Science has also fallen short with its ideas about origins of the Universe, of life, and of modern man. In fact, a theology has even emerged around the ideas of Darwinian Evolution for some pseudoscientists who want to dispel any ideas of a universal intelligence and consciousness. But as science moves ever deeper into quantum knowledge, science, and discoveries, it is moving ever closer to the dynamic Kabbalah.

The Bible has been translated, interpreted and misinterpreted for centuries. So it is not the Bible that is at fault. The Bible, in its original language, has as its foundation, the principles of Kabbalah. Once you begin to see the principles of Kabbalah within the Bible, your study of the Bible will take on ever-deeper meaning. As you begin to learn, and begin to understand some of the principles of Kabbalah, your level of comprehension and understanding will rise far above those around you. You will begin to learn that you can exert more control over many of the events in your life. You may even learn that you can influence the environment around you as it relates to you. And you may even find yourself in closer communion with "God."

A final word of instruction before we begin. Learning is much more than just stuffing information into your brain. To really learn something, you must also put it into use. This is especially true in a study of the dynamic Kabbalah. In other words . . .

If you don't use it, you will lose it!

Why study Dynamic Kabbalah and the Bible?

1. Principles of Kabbalah are revealed only to that person who strongly desires to learn.
2. Change is necessary for spiritual advancement.
3. Chaos is a condition for something new to emerge.
4. You must change your perception before you can change yourself.
5. Your thought forms reflect your attitudes, which in turn affect your relationship with others.
6. Each cell of your body contains all of your memories.
7. The foundation of all ancient religions is based on the Principles of Kabbalah.

Chapter 2

THE NEW WORLD

"Then the angel showed me the river of the water of life, as clear as crystal, flowing from the throne of God and of the Lamb, down the middle of the great street of the city. On each side of the river stood the tree of life, bearing twelve crops of fruit, yielding its fruit every month. And the leaves of the tree are for the healing of the nations." Revelations 22:1-2.

The Tree of Life is mentioned in only 2 places in the Bible. It is mentioned in Genesis the second chapter of the Bible, which explains Adam and Eve in the Garden of Eden with the Tree of Life. The only other place in the Bible where the *Tree of Life* is mentioned is in the last chapter of the Bible, which is quoted above. It is not an accident that the Bible opens and closes with the disclosure of the Tree of Life.

In the Genesis account it is simply stated without any explanation that the Tree of Life grants eternal life. Being cut off from the Tree of Life begins the process of death. This is all that we learn from a casual reading of the Genesis account.

In Revelation 22 quoted above, we find that a river of water of life symbolizes eternal life, and it flows from the throne of God and of the Lamb. On either side of the river is a street that contains a tree of life. The two trees and the water of life all contribute to the life-giving fruit. The leaves of the Tree of Life can heal even the nations.

Is this just an idyllic picture or does it have real meaning for you and me? We need to go to the Kabbalah to learn more about the Tree of Life and about how it relates to you and me.

The written Kabbalah was actually constructed as a commentary on the Pentateuch or the 5 books written by Moses. The 5 books of Moses, (1st 5 books of the Old Testament) are called The Law. The Bible has fallen out of favor with most people today because so little of it makes any sense to them. In reality, the Bible was written so as to obscure the true wisdom and understanding from everyone except a true Kabbalist. So the written Kabbalah was devised to reveal the foundation of the Bible, or The Law.

A Kabbalist wrote: **Woe to the man who sees nothing but simple stories and ordinary words in the Law! For were this so we could even nowadays frame a law which would deserve higher praise . . . . But it is not so; every word of the Law holds an exalted meaning and a sublime mystery.**

The Kabbalist understood, for example, that the earth was not created in 6 days as is expressed by the account in Genesis. But the Kabbalist did not want to put himself above the historic accounts in the scriptures or its positive precepts, so he searched for its deeper meaning.

In the words of a Kabbalist, **"If the Law consisted of nothing but ordinary words and recitals, like the words of Esau, Hagar, Laban, Balaam's ass and Balaam himself, why should it have been called the Law of truth, the perfect Law, the faithful testimony of God? Why should the wise man deem it more precious than gold and pearls? But it is not so. Every word conceals a most elevated meaning; every recital contains more than the events it seems to contain. And this higher and more holy Law is the true Law."**

There were people who were able to discern a deeper insight into the Law even though they knew nothing about the Kabbalah. The Kabbalah was first put into words during the Middle Ages even though it was taught with the greatest of secrecy before then. A few of the concepts of Kabbalah were taught in the Mystery Religions of the ancient Babylonians, the Greeks and the Romans even though the Kabbalah, itself, may have been unknown to them.

One of the early Catholic Church fathers (Origen) wrote, **"Were we obliged to hold to the letter of the Law and to understand what is written in the laws as the Jews and the people understand it, I should blush to proclaim that it is God who gave us such laws; I should find more grandeur and more reason in the laws of man as, for instance, in the laws of Athens, Rome, or Lacedemonia . . . . What sensible man, pray, could be made to believe that the first, second, and third days of Creation, where morning and evening are mentioned, could exist without sun, moon, and stars; that on the first day there was not even a sky; where will we find a mind so limited as to believe that God devoted Himself like a farmer to the planting of trees in the Garden of Eden, situated to the East; that one of the trees was the tree of life, and that another could give knowledge of good and evil? I think that no one will hesitate to regard these as facades behind which mysteries are hidden."**

Origen, (Adamantins Origenes, born in 185, died in 254) explains that there are 3 levels of understanding among people. He labeled them as having an understanding of an historical meaning, a moral meaning, or a mystical meaning. An *historical meaning* he likens to the *body*. It is the material view or understanding which accepts the meaning of a given text at face value. A moral meaning is a view that gives a moral lesson to the story or text. The story, or allegory, is used to illustrate the moral lesson. He likens the *moral meaning* to the **soul**. A mystical meaning is one that is hidden by the text. It is the foundational meaning on which the text is based. It is the reason as to why the text is as it is. A *mystical meaning* corresponds to the *spirit*. The Kabbalists also delineated three levels of understanding. The *simple-minded* could see only the Law, itself, and nothing beyond it. The *well informed* could see Law, as well as the context in which the Law was placed. The *wise* could fathom the whole foundation on which the Law and its context were placed. And that foundation was, to them, the *true Law*.

YAHSHUA (Jesus) understood the Kabbalah and demonstrated this knowledge when he was asked: **"Teacher, which is the greatest commandment in the Law?"** He replied, **"Love the Lord your God with all your heart, with all your soul, and with all your mind. This is the first and greatest commandment. And the second is like it: Love your neighbor as yourself. All the Law and the Prophets hang on these two commandments."**

It is commonly taught that the Kabbalah began with Abraham. Abraham is the central figure in 3 of the major religions on earth. He never created a religion while he was alive, but 3 religions today consider him to be one of their founding fathers. These religions are Judaism, Christianity, and Islam.

Abraham was called the "father of many nations." But some people have wondered how this could be since the only nation they have ever heard of was the nation of Israel. Abraham was the progenitor of Israel through his son Isaac. Abraham was a man of power and influence even though it was only hinted at in the Bible. The Jewish race and religion consider Abraham to be their father even today. Present day Palestinians, Jordanians, and several other Arab groups consider Abraham to be their father too. They trace their lineage through Abraham's son Ishmael, and through Abraham's six sons with his second wife Keturah. This is why Abraham is important even in the nations of Islam.

Christians consider Abraham to be important because he was mentioned in the New Testament as an example of one who had faith in God.

James 2:23 **"And the scripture was fulfilled that says, Abraham believed God and it was credited to him for righteousness, and he was called God's friend."**

How could anyone human be God's friend? Why is this

statement so important? It is of extreme importance because it points right back to the Kabbalah.

In the Old Testament, II Chronicles chapter 20, King Jehoshaphat actually invokes the name of Abraham as God's friend for protection: **"O Lord, God of our fathers, are you not the God who is in heaven? You rule over all the kingdoms of the nations. Power and might are in your hand and none can withstand you. O our God, did you not drive out the inhabitants of this land before your people Israel and give it forever to the descendants of Abraham your friend?"**

The king continues to explain that 3 major armies are coming together to drive them out of the land of Israel. Israel is unequipped and unable to mount a defense. God answers through the mouth of someone in the congregation. He instructs the king to take the army and have them overlook a certain valley. God, himself, would destroy the 3 armies.

When Israel's army reached the appointed place overlooking the valley, all they could see for miles around were dead bodies of soldiers.

What had happened?

It seems that as the 3 armies reached their rendezvous, 2 of the armies attacked and destroyed the third army. The 2 remaining armies then turned on each other until not one man was left standing.

Why was this incident important?

When God called Abraham his friend, it was an honor that had apparently been bestowed on no other human being. And when a prayer was offered to God as a reminder of that ancient friendship, it was immediately honored.

Abraham was the first person to figure out the mathematical formula on which the knowledge of Kabbalah was based. The most ancient text of Kabbalah put it like this: **"And when the patriarch Abraham had considered, examined, fathomed, and grasped the meaning of all these things, the Master of the Universe manifested Himself to him and called him His friend, and entered into an eternal covenant with him and his posterity.**

Abraham then believed in God, and that was reckoned unto him as an act of justice; and the glory of God was called upon him; for it is to Abraham that the verse applies: 'I have known thee before I formed thee in the womb of thy mother.'"

This quotation from the Kabbalah contains a profound amount of knowledge, which we will not go into at this time. However, notice that the Master of the Universe (God, if you will) manifested himself in such a way as to be seen by Abraham. He apparently even carried on a conversation with Abraham during which he called him his friend. All of this because Abraham had discovered the relationship between God, the Universe, and humanity. Abraham had somehow gained an understanding of the origin of the Universe and of mankind, that was thousands of years ahead of his contemporaries. Some of that knowledge can be made available to you today.

It is said that Abraham wrote his discoveries in two books. It is from these 2 books that the principles Kabbalah were originally formulated. These books are the *Sefer Yetzirah* and the *Zohar*.

The more ancient of the two, and by far the largest, is the *Sefer Yetzirah*, which translates to the "Book of Formation." It contains a cosmology, or a system of physics, which is comparable to our current system of quantum physics. The Book of Formation also explains how the Universe was created using 32 paths of Wisdom. This is also the basis for "numerology." The 32 paths of Wisdom describe the *Tree of Life*, which we will later study in detail.

It is an accepted fact that the entire universe is based upon numbers. It was Pythagoras who said, "Nature Geometrizes." Carl Jung believed that numbers preexisted human consciousness. He believed that numbers were not invented but were discovered, because they always existed. This might be easier to understand when you realize that all of the information on the Internet, and in your computer, are stored and accessed only by use of numbers. These numbers are then translated into letters and words for you to read, pictures for you to see, and voices, music and sounds for you to hear.

The second book, and also the most important of the two, is

the *Zohar*, which means "Brightness." It deals with God, the spirit world, and the human soul. This book reveals the purpose for the Universe and for the human race. This can best be expressed by the following Kabbalistic statement.

"{God} had already created and destroyed several worlds before He decided to create the world we live in, and when that last act was about to be accomplished, all the creatures of the universe and everything that was about to be in the world (at whatever time they were to exist) were present before God in their real form before becoming a part of the Universe. It is in this sense that we should understand the words of Ecclesiastes: 'That which is hath been long ago, and that which is to be hath already been.' The entire lower world was created in the likeness of the higher world. All that exists in the higher world appears like an image in this lower world; yet all this is but One.

This may sound like double talk, but it really makes sense if you understand the real meaning. The Creationist view of the formation of the Universe and all that it contains is that God began with nothing. Then, using his tremendous energy, he formed the sun, the moon, and the stars, and then set in motion the laws to govern them. After creating the earth in the same manner, he took soil from the earth and from it He formed all of the variety that exists on the planet. He also formed the first man and woman from the dust of the earth. Then He set in motion all of the laws that govern nature. But man received a human nature, and became aberrant. This is *not really* how the Universe came into being.

The Theory of Evolution, which is taught in the schools and colleges all across America today, presents an alternative to the Creationist Theory. Evolution begins with a Big Bang. In this way the Universe springs into being out of nothing. Evolutionists do not usually deal with what happened before the Big Bang, or where the energy stored in the Big Bang originally came from. Then using the laws of physics and chemistry they deduced how the Universe cooled down into physical matter and evolved into what we can see today. Continuing to use the laws of chemistry and molecular biology, they advanced the Theory that atoms formed into molecules and molecules formed amino acids from which single celled life began. They have not yet figured out how non-life suddenly became alive, but once becoming alive the single cell spread like a virus! Eventually more complex forms of life evolved from the simpler ones until the highest form of life, Man, evolved. No attempt by strict Evolutionists has been made to explain where all of the laws (of nature, physics, chemistry, biochemistry, molecular biology, gravity, atomic forces, electro-magnetics, electricity, and life) originated, and what causes them to continue to operate and interact continuously and without fail. This is not really where life or where man came from, either.

Then How Did It Happen?

Have you ever seen a magnet placed on a table and then iron filings sprinkled around it? What happens? You are correct! The iron filings form a design around the magnet. No matter how you sprinkle the iron filings or how many magnets you have, the basic designs made by the iron filings will all be similar. Do you know why?

The magnet emits a magnetic force field around itself. Iron filings are attracted to that force field and place themselves within that magnetic field structure. The iron filings have clothed that force field. The magnetic field around a magnet has always existed, but you cannot see it until the iron filings clothe it. The real form, which is "with God," is the magnetic force field. The iron filings create the image, in the lower world. This is called *Emanation*, and is how the material world came into being!

The Western Worldview of the Universe and all that it contains begins with the material substance and then tries to explain what it is and how it behaves.

The view from the Kabbalah is just the opposite. It begins with the explanation of what the Universe is and how it behaves, then shows how it came into material form. Following is another quotation from a Kabbalist about the emanation of the Universe that goes even deeper than the previous Kabbalistic statement:

"Before God manifested Himself when all things were still

hidden in Him, He was the least known of all the unknowns. In that state He had no other name than that which expresses interrogation. He began by forming an imperceptible point; that was His own thought. With this thought He then began to construct a mysterious and holy form; finally He covered it with a rich radiant garment—that is to say, the Universe, whose name necessarily enters into the name of God.

As we learned earlier, thoughts are pure energy. With our thoughts we can build thought forms or thought structures. These are energy structures much like the magnetic force field, only thoughts operate on a much higher level of vibration. Once the idea of the Universe was completed and perfected in His thoughts, He clothed it, or brought it into being.

We can do the same things with our own minds. Every time you conceive of creating something in your mind, and then bring it into being, you have constructed a mysterious form and then covered it. Whether you build a bookshelf, or bake a cake, you first form the thought, and the picture, in your own mind and when it stands before you finished, you have clothed that thought. Some people have powers of concentration of thought sufficient to accomplish this simply though the intense focus their own minds.

The power of thought or of creativity does not exist in the animal world. You see, man is not the pinnacle of evolution. The human race is a special creation or emanation and is designed for a specific purpose. It is the possession of a mind that separates mankind from the Animal Kingdom. And the mind is much more than just the product of human brain activity.

The *Zohar* reveals, **"Man is both the summary and the highest expression of Creation; hence he was not created until the sixth day. As soon as man appeared everything was completed, the higher world as well as the lower, for all is summed up in man, he unites all form."**

The higher world has existed long before the material world came into being, as in the example of the magnetic fields and the iron filings. The material world was not required until humanity needed a more solid form, which necessitated a material universe.

That is why man unites the higher and the lower worlds. This may sound like the height of conceit but in time you will come to understand why this is so.

The *Zohar* continues, **"Do not think that man is but flesh, skin, bones and veins; far from it! What really makes man is his soul; and the things we call skin, flesh, bones and veins are but a garment, a cloak; they do not constitute man. When man departs this earth he divests himself of all the veils that conceal him. Yet, the different parts of the body conform to the secrets of the supreme wisdom . . . . Celestial Adam is as spiritual as terrestrial man, and everything happens below as it does on high. Therefore it is written in scriptures: 'And God created man in His image.'"**

The Kabbalists have shown that a spiritual, or an energy-information world exists, which is the foundation of the material world. This higher energy world has pre-existed the present material world in all of its forms. Since man is the bridge between the higher energy-information world and the lower material world, it stands to reason that man also has pre-existed his appearance on the material world. When I speak of man, of course, I mean his/her soul.

Have you ever wondered why people are driven to accomplish something with their lives? Why are you reading this book instead of doing something "fun"? The position of your soul as the bridge between the energy-information world and the material world creates a need for the desire to improve the Universe. For some people that translates into the desire to improve their own environment or neighborhood. For others it might mean improving the earth on which we live. And for still others it means venturing out into space to explore the Universe. This is how the soul manages to discover who it is, why it is here, where it came from, and how it will return. The soul also desires to discover the path it is traveling and how to know its ultimate destination. It is only the few who make those discoveries genuine.

The earth is an educational system for the soul. Souls come into the material world and are born into a body. The few years that the soul spends on earth are an educational experience. Some souls learn more than others do, and some learn nothing at all. Much of the learning process involves interaction with other people. And when it is all over, the body is shed and the soul returns to the energy world. We will learn more about this as we study . . . .

The Tree of Life.

The New World

1. The Tree of Life is mentioned two times in the Bible; at the beginning of Genesis, and at the end of the book of Revelation.
2. There is more to the Law (Old Testament) than the story and the words convey.
3. The words of the Law conceal a much higher meaning.
4. The words of the Law give an historical meaning, reveal a moral meaning and conceal a mystical (spiritual) meaning.
5. Abraham discovered the Kabbalah and revealed in in 2 books; *Sefer Yetzirah* and *Zohar*.
6. The lower (material) world is an exact image of a pre-existent higher world.
7. The lower world clothes the higher world like a fine garment.
8. The soul of Man unites the upper and the lower worlds.

Chapter 3

THE EMANATION OF THE UNIVERSE

Where did the Universe come from and how did it get here?

We all have preconceptions about how the Universe came into being, and our beliefs flavor practically everything we understand about the Universe, our personal origins, and our purpose in this vast Universe. Some people were taught the theory of a literal 7-day creation as is described in the Genesis account in the Bible. In school and in college you were taught the *Theory of Evolution*, which depicts the Genesis account as a rather loose sequence of events indicating a series of evolutionary steps rather than a sequence of daily creative acts. Your present belief is probably a combination of these two factors.

As we have learned in the previous chapter the Kabbalistic view of the origin of the Universe is called *"emanation."* This is a concept that means "to flow out from," or to "proceed from a source of origin." Emanation contrasts with Creation in that God did not use his hands to create something out of nothing. Emanation also contrasts with Evolution in that the Universe did not pop out of nothingness, or emptiness. Dynamic Kabbalah explains that the beginning of life on earth was not an unplanned accident.

An ancient sacred book of India, the *Rig-Veda*, gives the following accurate explanation:

In the beginning there was neither existence nor nonexistence.

All of this world was unmanifested energy.

The one breathed without breath, by its own power.

Nothing else was there!

A Kabbalist has stated that the name for God and the name for the Universe in the Hebrew language are one and the same. They are identical. When God revealed "Himself" to Moses at the burning bush, He said his name was *I AM THAT I AM.* In English it means "I exist so that I may exist." Or, "I exist by my own power." As you can see, both the *Rig-Veda* and the *Kabbalah* agree that there was a source of unmanifested energy for the beginning of the Universe.

The Apostle Paul was both a mystic and a Kabbalist. He stated the same concept with the following words in Romans 1:20:

The invisible things of him from the creation of the world are clearly seen, being understood by the things that are made, even His eternal power and Godhead.

What Paul is actually saying is that everything in the material Universe has an invisible counterpart. Or to put it yet another way, there is a spiritual or energy pattern in the Universe for everything that you can taste, touch, smell, see or hear. You can even understand the nature of power and rulership of the Universe simply by studying the power structures and all of the forces of the material world.

To begin our study of the Emanation of the Universe we can look to the first chapter of Genesis.

Genesis 1:1, "In the beginning God created the heaven and the earth."

This is a statement about the beginning of the Universe. Time did not begin until the material Universe came into being. But before the manifestation of the Universe there existed primal, pure, unmanifested energy. The invisible Universe was already there just awaiting its manifestation.

Genesis 1:2, "And the earth was formless and void, and darkness was upon the face of the deep. And the spirit of God moved upon the face of the waters."

The unmanifested energy had no form. It existed only as the laws of physics, chemistry, gravity, and atomic and molecular forces. The "deep" was a gigantic container, or force field that contained the potential energy. The formless energy was darkness because it emitted no light. The "waters" were that unmanifested potential energy. The spirit of God began to "flow" into the unmanifested energy and with a burst of light, the *emanation* of the material universe began!

Genesis 1:3-5, "And God said, let there be light and there was light. And God saw the light, that it was good, and God divided the light from the darkness. And God called the light Day, and the darkness he called Night. And the evening and the morning were the first day."

These first 5 verses of Genesis are a revelation of what occurred *before* the Universe materialized in its present form. It explains what happened before time began. It explains the cause of the Big Bang, which resulted in the formation of this Universe. The *light* and the *darkness* therefore, refer to the formation of a principle of creative forces, which were required to bring the Universe into being. Let me now explain the emanation of the Universe from the viewpoint of the dynamic Kabbalah.

Our present Universe originally began with the unmanifested First Principle. The unmanifested First Principle, which we will call the *FIRST CAUSE,* became, or existed as a great, gigantic thought form. This thought form contained within it every thought and idea that would ever become manifested in the Universe. This thought form also contained within it all of the laws of the Universe that would be needed to manifest and maintain the Universe and all that it would ever contain. This thought form was a perfect blend of completely undifferentiated thought. It was/is one perfect thought, not a collection of thoughts as we experience. This was/is the Eastern concept of Nirvana. In the Kabbalah it is called *AIN*.

The FIRST CAUSE has also been called *White Light*. This is not the kind of light that the human eye can see. It is a very highly intensive form of energy-information. As we have already learned, *thought* is also a form of energy. In fact it is the most powerful energy in the Universe, when properly focused. *White Light* vibrates at such an intensively high rate of vibration that any contact with the material universe would immediately disintegrate it into nothingness. Therefore, the first emanation had to be a container, or a force field of sorts, to separate anything created from the pure White Light. This container has often been called the "cosmic egg." Ancient religions have also referred to it as the "cosmic womb." The Kabbalistic term is *AIN SOPH*. This container embodied a vacuum of darkness in the midst of the surrounding White Light and would forever separate whatever was within the container from the destructive power of the pure White Light.

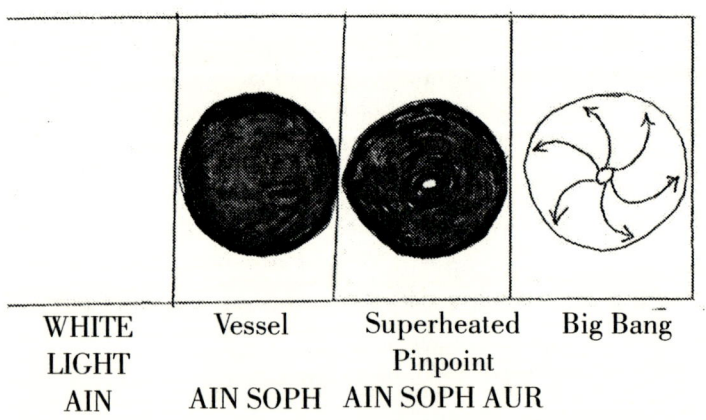

WHITE	Vessel	Superheated	Big Bang
LIGHT		Pinpoint	
AIN	AIN SOPH	AIN SOPH AUR	

cosmic vessel

Within the container there were now two potential forces, light and darkness. Light contained a *positive* charge and darkness had a *negative* charge. Light symbolized the masculine principle and darkness (cosmic womb) symbolized the feminine principle. The FIRST CAUSE had now become Male and Female. The male and female components were necessary for the emanation of a material universe. This is what the light and darkness of the Biblical first day represents. Ejaculation of positively charged *White Light* into the negatively charged dark container (womb) created all of the energy forms, chemical, physical, atomic, and gravitational forces out of which the Universe would be formed. The negative forces of the container walls compressed the positive white light into a tiny pinpoint at the center of the container. The contraction of the energy into a tiny point caused it to superheat and then explode. The explosion is what the scientists call the Big Bang. The Kabbalah refers to this as the *climax* of the first sex act.

The words of the Bible conceal Kabbalistic ideas far beyond the meaning of the words themselves. In the Hebrew language each letter of the alphabet is a word having its own meaning. So using all of the letters in the words of a single sentence as words themselves, you gain a much broader understanding of that sentence. Migene Gonzalez-Whippler, in *A Kabbalah For the Modern World*, has provided the following Kabbalistic interpretation of the words, **"And God said, 'let there be light and there was light.'"**

Interpreting the Hebrew letters from the words above, she writes; **"The Divine Fertilizing Agent projected in continuity the dual principle of life and death into the cosmic womb, and started the gestation process which brought forth the spiritual essence of the Created Universe. The Manifested Cosmic Principle was then surrounded by continuous existence. The essence of spirit was fertilized and expressed in universal manifestation through the cosmic copulation between male and female Cosmic Principles. This brought forth the Creation of the Physical Universe in the form of a vast explosion of light."**

The Big Bang marks the beginning of time in our Universe. The Emanation of the Universe may be calculated in fractions of a second, while the age of the Universe is thought to be in the billions of years. The material Universe is still contained within the cosmic container to protect it from the intensity of the pure white light. Time exists nowhere except in the material world. Past, Present and Future are all simultaneous outside of the manifested Universe.

The First Cause, or Great Cosmic Thought was all of the knowledge that ever existed or ever will exist. But the raw knowledge was useless until it was differentiated and categorized. The first category had to be the complete idea of duality. A material universe would not be possible without the dual principle of life and death. Decay at a molecular level, where gases cool down to liquids and solids, is necessary for the early formative process of the Universe. Suns, star systems and planets could not have formed without the ongoing process of decay. On an Ecological scale, decay and death are also necessary to maintain an ongoing balance. Without death and decay the food chain would be impossible with no method of continuing to sustain new generations of living creatures and plant life.

The principle of *continuous existence* is necessary in a manifested Universe because it is the *reflection* of conditions outside of the manifested Universe. In other words, life outside of the material existence is simultaneous and eternal. Life within the material realm is continuous and must be constantly renewed, revitalized, reconstituted and reproduced. This is why the first Emanation had to be Light and Darkness, positive and negative, male and female. The material Universe began its existence as a result of a sexual union, and its continuous existence will always be dependent on a wide variety of various productive sexual unions.

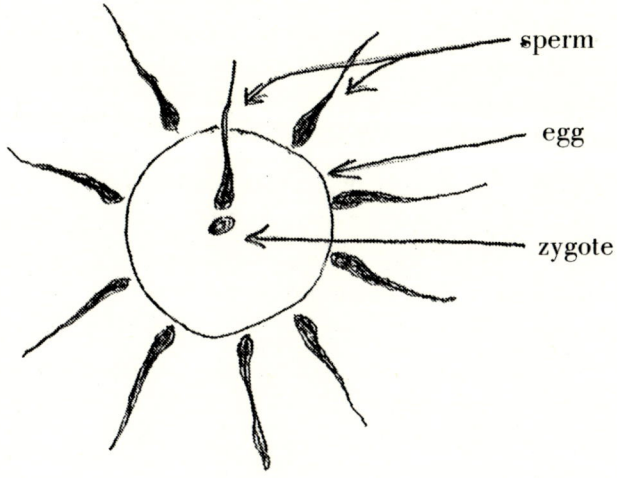

cosmic egg

All of existence outside of the material realm is called the *ENDLESS WORLD*. This is the world of pure spirit or White Light, where everything past, present, and future is perpetually simultaneous. It might seem logical to us that the First Cause would need lots of time to develop a well-balanced Universe. How long would it take to build something that would continue to function flawlessly for billions and billions of years? In the Endless World there is no time, so everything happens instantaneously and simultaneously.

The *SHEKINAH* is the active intelligence of creation. She is the embodiment of revelation as it relates to the enlightenment of the human mind. Without the Shekinah no creative activity would have taken place. She is the container, the cosmic womb. She is the feminine aspect of God. *YAHWEH* is the self-existent one. He came into existence by himself for the purpose of bringing all else into existence. By creating the Shekinah as his opposite, He brought himself into existence as well. Before this event there was no *YAHWEH*. There was only the undefinable White Light.

Yahweh became the male counterpart to the female Shekinah. The union of the two created all of the necessary components required for the development of life forms throughout the entire Universe. The pattern of the First Cause is replicated each time a single cell divides and clones itself. Moreover, the cells of your own body follow that same pattern. Even the most complex life forms follow this same positive-negative pattern of sexual reproduction.

You have just seen a glimpse of how the Universe came into being. The Universe is like a Gigantic LIFE FORM. It is composed of billions and billions of gigantic molecules and atoms with swirling electrons. We call them galaxies, star systems, planets and moons. This has been called The Heavenly man, or *Adam Kadmon.*

The *Tree of Life* is a symbol of the Adam Kadmon. It symbolizes everything that constitutes the makeup and composition of mankind. The Tree of Life is also a symbol of Man and Woman, as a composite unit. In the Kabbalah this unit is referred to as MAN. The Adam Kadmon is the macrocosm of all of the elements of the substance of humanity, including material, mental, moral and spiritual. Adam Kadmon is the Second Adam that is mentioned in the Bible by the Apostle Paul. He explains that the Second Adam is the animating or enlivening spirit of Man.

All of the descendants of Adam are also represented by the Tree of Life. They are the microcosm of the Tree of Life. They are the reflection of the macrocosm or the Adam Kadmon. Another way of viewing this is that all of the elements and forces that comprise the Universe are also available in MAN. These forces include powers, energies, levels of consciousness and awareness, and even life itself. The Tree of Life also organizes pathways through which all of these elements and forces can be accessed and used. This means that Man is a miniature model of the Universe! As you learn more about the Tree of Life, you will also learn how to use it to . . .

Change your life.

The Emanation of the Universe

1. In the beginning there was neither existence nor nonexistence.
2. The invisible Universe was just there awaiting its manifestation.
3. The first Emanation was the duality of Light and Darkness.
4. Light and Darkness, which is Positive and Negative energy, also defines male and female.
5. The Big Bang of the emanation of the Universe was the climax of the first Cosmic sex act.
6. YAHWEH is the masculine and SHEKINAH is the feminine aspect of GOD.
7. The Universe is like a gigantic life form, and is called the ADAM KADMON.
8. The Tree of Life, a symbol of Adam Kadmon, and of Mankind, organizes pathways to all of the forces and powers of higher Consciousness within Man.

Chapter 4

SOUL IS THE BRIDGE BETWEEN "GOD" AND MAN

The Endless World lies beyond the grasp of our finite senses. Yet it also exists within our own physical self. This may seem like a contradiction since it has been stated that any contact with White Light would disintegrate the material world. But provisions were made to overcome this problem.

The only way I know how to describe the Endless World is that it is energy at an extremely high rate of vibration. This energy is the sum total of all conceivable knowledge that is in a quiet state of undifferentiation. It is the Nirvana of the Buddhist and Hindu. It is the complete absorption of ego and individuality into the Universal Spirit. It is the state of complete and total Peace. But these are only words and in no way can convey to our finite minds the Endless World.

Having said all of that, it is the Endless World from which we came. And it is the *soul* that bridges the gap between the Endless World and the mind of Man. It is also the soul that gives the material Universe its purpose and meaning.

The Endless World wanted to share its wealth of knowledge, understanding and wisdom. But to do so would require differentiation. It would require the absence of peace and knowledge. It would require a container of separation. This container was the *Cosmic Womb*, or the AIN SOPH. Along with everything else that went into the container were also the Sparks of Life.

The Sparks of Life were a part of YAHWEH that would live in the material world to experience all of its wonders and horrors. The sparks of life would provide the Endless World with the opportunity to share. The sparks of life might be likened to an individual cell of Endless World DNA. In other words, they contained all of the knowledge of the Endless World. So in order to share knowledge with the sparks, the sparks first had to be empty, or devoid of knowledge. This was accomplished when the sparks allowed themselves to be emptied of all knowledge. An empty mind creates an extreme desire to be filled. Therefore, that strong "desire to receive" prompted the desire to share. The Endless World then shared its knowledge with each spark until the sparks were full, content and at peace.

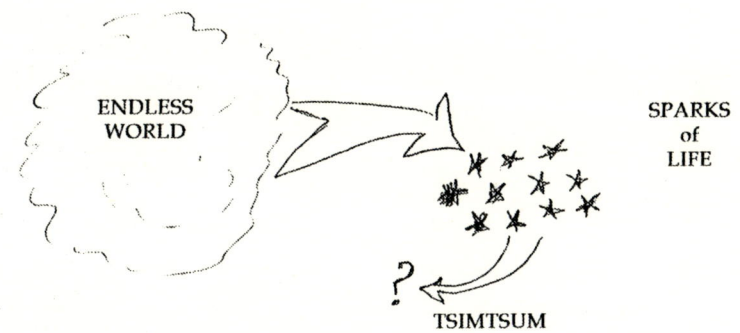

Sparks

The sparks desired to become an equal part of the Endless World by the desire to share its knowledge with others. But there were no others to share with because all were full and content.

This led to a condition of restriction called *Tsimtsum*. The sparks desired to share with the Endless World but *it* also had no need to receive anything because *it* was complete in itself and needed nothing. So the sparks stopped emptying themselves, which ended their receiving of light from others, and their own lights began to turn dim. There was nothing they could give in return for all that they had received, and they felt ashamed. This is called *Bread of Shame*. They became mere shadows of their former selves.

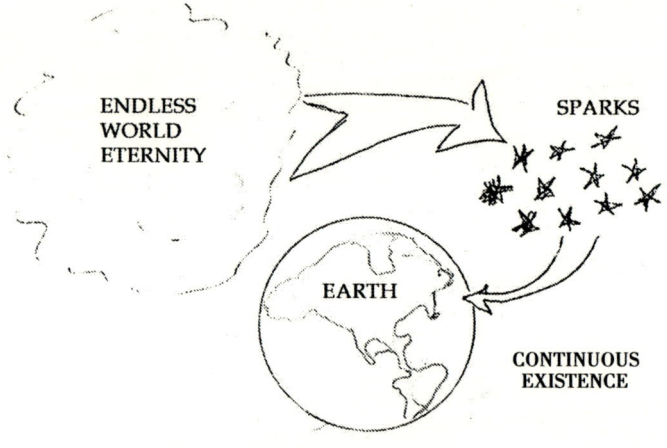

Desire to Share

The Kabbalah teaches that there are two forms of desire. There is the *desire to receive for self alone*. This is manifested by the desire to accumulate things for the self with little thought of giving anything in return. This is also representative of those who take and steal what is not theirs to use for their own personal gain. The light of these souls glows very dim.

The second form of desire is the *desire to receive to share with others*. This is the desire that led to the Bread of Shame because there was no one left with whom to share. Once these souls find someone with whom to share, their lights shine brightly again.

The material Universe of death and decay provides the perfect conditions necessary for the continuous opportunity to share and thereby begin to remove the Bread of Shame. This is why the Bible says in Ephesians 1:4, **"(God) . . . had chosen us in him before the 'foundation of the world' that we should be holy and without blame before him in love."**

This is in reference to the sparks of life and their condition that existed before the Emanation of the material Universe. Our material existence provides us with the opportunities to change our previous condition. This is why we exist in material form. The desire to receive while we are in the material plane of existence manifests itself as a continual search for satisfaction. Our internal sparks are trying to reclaim what they once had.

Your spark silently asks you the question, "Isn't there more to life than this?" You go to school to prepare yourself to get a job or an occupation so you can get married and raise a family. You busily provide for a place to live, cars to drive and try to accumulate as many things as possible for yourself and your family. You also try to provide for your retirement and your last years of life. But the silent question still persists. What else should I do with my life?

How do the internal sparks of life exist in this material world?

When the sparks of life come into the material world they are contained within a vehicle of protection. This is similar to the space shuttle taking people into the hostile regions of space. The shuttle protects the astronauts from certain death and also communicates with all of the control and guidance systems back at mission control. When the mission is finished the shuttle returns the astronauts back to the safety of "home" with all of the new knowledge and wisdom they acquired while on the mission.

The vehicle of protection for a spark of life when it enters the hostile environment of a decaying Universe, is called the *Soul*. The Soul protects the spark as well as provides the entire guidance and communication system while it is in the material realm. That is to say it is the soul that communicates with the material self. The spark will communicate back to the Endless World but not in any way that we would understand. The soul is a communication system designed to communicate and translate between the language of human thought and the energy system of the Universe. In order to understand what the Soul is, we must first learn about the Four Worlds of Emanation.

The Kabbalah teaches that there are *Four Worlds* that make up and sustain the entire Universe. When Abraham discovered the true nature of God, he stopped calling him God (Elohim). Abraham began to use a term similar to *Divine Thought*. This is the same as using the terms *White Light* or *First Cause*. It all means the same thing. The first of the four Worlds is Divine Thought, or The World of Emanations.

World of Emanation is the world of *Archetypes*. This is the world of pure thought. There are no words and no pictures. There is only the pure energy of Divine Thought. Everything that exists, has ever existed, and will ever exist, has its beginning here as a pure and perfect thought. The energy level is so intense that it is beyond the level of human thought.

World of Creation is the world of *Ideals*. This is where ideas begin to take shape. The interplay of positive and negative energy begins to energize the ideals into thought forms and patterns that can be captured by the astute mind. This is the mental plane of creativity, and is where Wisdom and Understanding emanate. The energy levels are less intense than in the higher World of Emanations. The World of Emanations and the World of Creation are the abode of a manifestation of the soul called *Neshamah*.

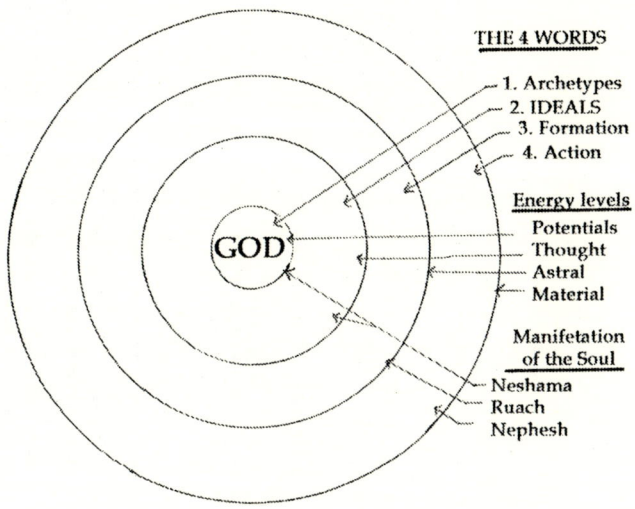

THE 4 WORDS

1. Archetypes
2. IDEALS
3. Formation
4. Action

Energy levels

Potentials
Thought
Astral
Material

Manifetation
of the Soul

Neshama
Ruach
Nephesh

GOD

4 Worlds

World of Formation is the world where shape and substance are added to the thought forms. This is also called the Astral World or Astral Plane of existence. The energy vibration levels are much lower here than in the previously mentioned higher worlds, and forms and shapes of things can actually be seen by sensitive people. This world is composed of a very subtle electrical energy that is in a constantly moving fluid state. This is the substratum on which the material world is molded or formed. Every material object has an astral form, which can be seen. However, there are many astral forms that do not have a material manifestation. The World of Formation is the abode of a manifestation of the soul called *Ruach*.

World of action is the world where everything comes together. This is the material world that is so familiar to us. The energy vibration rate is much lower in the material world so material objects can assume a very hard and stable form. Other materials can be liquid or even gaseous depending on their molecular density and vibrational rate. This is the world where we perceive life and death. The World of Action is the abode of a manifestation of the soul called *Nephesh*. The interplay of these three manifestations of the soul, Neshamah, Ruach, and Nephesh determine the substance of your character and your personality. Some Kabbalists say that there are more than three manifestations of the soul but it will make things much simpler to understand if we just use these three.

The Spark of Life in the human body occupies a space about the size of a grain of wheat. It can be evicted from the body through violent trauma, such as in a severe automobile accident, and when that occurs death is sudden and instantaneous. The Spark of Life is God within you. That is the meaning of the concept . . .

"To find God you must look within yourself."

You see, the Four Worlds of Emanation (*Emanation*, *Creation*, *Formation* and *Action*) are all within your physical body. The Spark of Life needs to be protected within each of those worlds and that is the reason for the manifestations of the soul. The three manifestations provide protection and interworld communication. Now let us look at these in greater detail.

NESHAMAH (nesh-aw-maw')

In the Bible and the Kabbalah, Neshamah is the term for the highest manifestation of the soul. All of the major religions have a similar concept. Neshamah is called the angry wind, and can mean vital breath and Divine Inspiration. It is also relative to the intellect. Neshamah is the higher self or Overself. It is also referred to as the Oversoul. Neshamah has access to all Wisdom and Understanding and can communicate through *intuition*, the will of God for your life.

The seat of communications for Neshemah is the brain. If the space shuttle were the soul for the Sparks of Life, Neshamah would be the Houston Mission Control Center, which coordinates communications for all space missions. It knows the mission of each space flight and evaluates the success of each mission both while it is in action and after it returns home.

Neshamah is the Transcendental Ego, or the real ego. It operates in the real or ideal world and transcends all levels of existence. Neshamah also evaluates the progress of the soul in the lower worlds, and tries to nudge the individual in the right direction whenever that can be done. Trying to discover the real ego is the essence of spiritual advancement.

Proverbs 20:27 says, **"the spirit (Neshamah) of man is the candle (light) of the Lord (Divine Thought) searching all the inward parts of the belly."** The belly refers to the lower portions of the soul that motivate the individual to do, or to be, whatever he or she is. Meditation is the process of communicating with this level of your soul, and allowing it to bring you closer to Divine Thought.

RUACH (roo'-ach)

In the Bible and Kabbalah Ruach is the term for your personality or ego. This is not the real ego, but the false ego. This is the **"I"** in your life. Ruach is the part of you that comes to mind when you think of who you are. Ruach is also what others consider when they think of you.

Ruach means the mind, spirit, wind, and breath. It was Ruach that was "breathed into Adam's nostrils that made him a Living Soul. The Greek word for Ruach is *"pneuma,"* which means a current of air, spirit, vitality, and mental disposition. When your spirits are low, or are uplifted, we mean Ruach. If you are mean spirited it is your Ruach that is to blame. Ruach also relates to your intellect because it is the center of your outward consciousness. Ruach gives you awareness and perception of thoughtforms, and assists you in molding thoughts into action.

Ruach is your moral world. Your concept of morality is fashioned here. Everything you believe is centered in Ruach and describes who you are. Memory is located here because it relates to intellect and storage of knowledge. Wisdom and Knowledge are collected here and processed for usefulness. Will is also developed here in Ruach because the power of will relates to your sense of morality and what you believe is right and wrong. Your own individualized consciousness comes into play as a part of Ruach because it is supposed to be used to investigate the nature of the Universe. Many folks spend a lot of time investigating that part of the nature of the Universe that relates to love, sex, and emotional security. Others investigate the relationship of power to money. Some investigate using telescopes and microscopes and electron accelerators, while yet others look into books, writings and ancient knowledge and wisdom. All of this and much, much more is under the influence and control of your Ego, Ruach.

The seat of communications for Ruach is the heart. In our space shuttle example, Ruach would relate to all of the scientific equipment on board used to perform the experiments and all related functions to the mission at hand. If the equipment or other conditions are unsatisfactory, a mission can turn out to be incomplete, or even aborted. At other times it could require a judgment call by Mission Control to determine a course of action.

Your conscience is the telephone used by Neshamah to contact Ruach about a matter that you may be neglecting. Neshamah might be trying to convince you to strengthen your will and commence a course of action. Since Ruach is centered in your heart, that is where you will feel the pain when Neshamah is calling you.

Ruach is what prevents the light of higher principles of Neshamah from shining through to your mind. Ruach is in charge of everything you think and believe and does not want to be challenged. When Ruach searches for "truth" it is usually looking for knowledge that supports and bolsters its present belief system. Ruach is centered in the heart. When someone says, 'follow your heart, not your head, they are speaking to your Ego, Ruach. When Ruach discovers something that challenges its core beliefs, it tends to discard or ignore it. When Neshamah phones you to say, "why don't you take a closer look at that concept or idea?" that is when you feel a twinge in your heart that grows ever stronger until you ultimately begin to investigate. Finally, when Ruach investigates, validates, and incorporates a new concept into the old belief system, a change in perception is achieved. Spiritual advancement is the process of revealing the false ego so as to encourage change.

Psalms 26:2 pleads, **"Examine me O Lord, and prove me; try my reins (kidneys) and my heart (seat of Ruach)."** When Ruach allows more knowledge and wisdom to filter down into its domain, you move a little closer to Divine Thought. Reins are a reference to the lower nature of the soul.

NEPHESH (neh'-fesh)

In the Bible and Kabbalah, *Nephesh* is the term for your lower nature. Nephesh means a breathing creature or animal. It is the vitality relating to life. It is the spirit that gives movement to the body and makes it alive. The Bible refers to Nephesh as a *Living Soul* when referring to Man, and *Living Creature* when referring to animals. The Greek word is *psuche*, which refers to spirit, or animal sentience. It is the animating principle common to both animals and man.

Nephesh is Man's lower nature and is the center for our animal instinct. Nephesh is the animal side of the soul, which comes into contact with the external Universe. Nephesh is the seat of our deepest desires and is what we use to explore the world of passion and sensuality.

Nephesh is an emanation of the soul that can actually be seen. It is composed of the subtle body and the Astral body. The subtle body is the electrical system that operates the material body of flesh and bones. Brain waves are part of the subtle body. Nerve impulses are also part of the subtle body. Nephesh is the clothing of the soul. The subtle body is clothed with a body of light called the Astral body. The Astral body is the foundation over which the material body is molded. If the lighting is just right and your eyes are relaxed, you can sometimes see the Astral body of another person.

The seat of communications for Nephesh is the kidney. When Nephesh is attempting to gain your attention it is called a gut feeling, which strikes you in the pit of your stomach. In our space shuttle example, Nephesh would be similar to the communication and guidance systems on board the shuttle. It would be the computer and communications system that physically operates the space shuttle itself. It basically operates by itself but can be overridden by the Astronauts or occasionally even by the Mission Control Center.

Nephesh is the link between Ruach and the human body. When someone dies a sudden violent death the soul is thrown from the body and does not realize that it is dead. Until the soul accepts its death it can be seen (or sensed) as a ghost. The ghost is the Nephesh or astral body that protects the spark of life until it leaves the material world. Nephesh can communicate with people, and can usually be instructed on how to escape the material existence and go on into the higher realms. Sometimes only the subtle body remains and it is detected as an electrical force field that has no sentience. It can be electrically neutralized. These are all effects of Nephesh.

Genesis 2:7 states, **"And the Lord God (Divine Thought) formed man from the dust of the ground and breathed into his nostrils the *Neshamah* of life, and Man became a living *Nephesh*."**

When referring to animals Nephesh is called Creatures, but when referring to man it is called Soul. It is Neshamah that makes Man much more than just a higher form of animal (Nephesh). Man has consciousness and is capable of understanding higher knowledge and wisdom. That capability to understand knowledge and wisdom also requires an evaluation as to how it is used.

Jeremiah 17:10 warns, **"I the Lord (through Neshamah) search the heart (seat of Ruach), I try the reins (seat of Nephesh), to give everyone according to his ways, according to the fruit of his doings."**

When the 3 emanations of the soul are working in harmony, spiritual advancement is taking place. Soul searching is necessary to cause a change of perception that leads to spiritual advancement.

Divine Thought is in all things. It is in every atom of every molecule of every substance on earth. Divine Thought is in your DNA coding. Divine Thought is in the air you breathe and even in the substance that makes up outer space. Even though Divine Thought is all around and within you, your brain cannot comprehend or understand it. The material brain cannot relate directly to Divine Thought. It is Neshamah that acts as translator of Divine Thought to your conscious mind, Ruach. Neshamah presents the elements of Divine Thought to your mind that would be beneficial to your spiritual advancement.

Spiritual advancement requires a change in your perception of yourself, which Ruach resists.

You are Ruach. That is to say, Ruach is the person you perceive yourself to be. Ruach defines your personality and your character to you. When Neshamah attempts to provide you with insights from Divine Thought, which challenge your present system of beliefs, your normal reaction will be to resist or block that insight into your own character and moral standing until you are ready to change the way you perceive of yourself. This is the function of Ruach so as to prevent you from going insane and completely losing your "personal identity." When you are ready to be shaken and to change, you will seek the counsel of Divine Thought through Neshamah with prayer, meditation, and a calming of your mental activities.

Added to a developing spiritual nature you also have an animal nature. This fact has been understood by people of all religions around the world for thousands of years but in this era of "political correctness," great attempts are being made to conceal this fact of life. This animal nature, also called human nature, is Nephesh. This is your tendency to be idle, prideful, crude, selfish, dishonest, lustful, resentful, hateful, tyrannical, destructive and vindictive. These tendencies, and more, can be expressed when Nephesh feels threatened. These tendencies of human nature help to define your standards of morality, your personality and your character. When you (Ruach) feel these emotions of Nephesh begin to well up inside, you will try to control them and align them with the image of your perception of yourself.

Nephesh is driven by the desire to receive for self, alone. Nephesh desires food, clothing, shelter, sex, love, respect, admiration and knowledge. Ruach tries to define how these things will be received and what will be offered in return. Neshamah counsels that whatever you receive should be used to share with others, selflessly, with no thought of return. Ruach tries to balance the desires of Nephesh against the higher ideals of Neshamah. The greatest desire is to find a mate with whom to share needs and desires throughout the rest of life.

When the sparks of life came out of the Endless World they were each a homogeneous unit just like Divine Thought. When they came into the material realm they were divided into male and female, just as Divine Thought became Yahweh and Shekinah. The male became a whole unit unto himself and the female became a whole unit unto herself. Together, the two souls are called *soul mates.*

This is the same pattern that was used to create Adam and Eve as described in Genesis 1:26-27, **"And (Divine Thought) said, 'Let us make man in our image, after our likeness . . . So (Divine Thought) created Man in his *own image*, in the image of (Divine Thought) created He *him*, male and female created He *them*."**

This shows that just as the Kabbalah teaches that souls began in an androgynous state, so the Bible teaches that Man began in an androgynous condition according to the image, or pattern, of Divine Thought.

Soul mates are destined to search for each other for eternity because of the affinity they have for one another. Most of the time they reincarnate on the earth at different times and in different lands so that it is very improbable they would ever meet and get back together. But the search is the motivation that keeps a soul yearning for a higher cause. This pattern is repeated in humanity in that the primary focus of most people as they approach adulthood is the search for a mate. Once a mate has been found the couple begins to investigate all that the world has to offer, whether in the sensuous world or the intellectual. But as we learned earlier, Neshamah is calling us to investigate the upper world from which we came.

Kabbalah teaches that Adam and Eve did not begin on the material plane but in a higher world: **"When our forefather Adam inhabited the Garden of Eden, he was clothed, as all are in heaven, with a garment made of the higher light. When he was driven from the Garden of Eden and was compelled to submit to the needs of this world, what happened? God, the scriptures tell us, made Adam and his wife tunics of skin and clothed them, for before this they had tunics of light, of that higher light used in Eden . . . The good actions accomplished by man on earth draw down on him a part of that higher light which shines in heaven. It is this light that serves him as a garment when he must enter into another world and appear before the Holy One . . . the soul has a different garment for each of the two worlds it must inhabit—one for the earthly world and one for the higher world."**

Beyond consciousness of a material world there are other levels of consciousness as well. Many people are aware of nothing more than their physical surroundings. But more and more people are beginning to awaken to the fact that there are other worlds of consciousness beyond the human senses. There is the etheric world of ghosts, apparitions and subtle electrical forces. You have an etheric body. There is also the astral world of fluid electrical energy, which is the foundation of everything material that exists. You also have an astral body. Your consciousness can exit the material body in the astral body and travel unencumbered anywhere. This often happens while you are asleep. If you have an extremely vivid dream of being somewhere, it may not have been a dream, but a real astral experience. The etheric and astral levels of consciousness are in the realm of Nephesh.

There is also the world of mental consciousness where mental activity and thought energy occurs. Thoughts are organized into thought forms, which is the foundation for everything that exists in the material plane. Learning to use and focus thoughts to manifest changes in the material plane can be learned at this level. This is the consciousness level of Ruach.

Rumi, a mystic and a poet once wrote:

> **There is someone who looks after us**
> **From behind the curtain**
> **In truth we are not here**
> **This is our shadow.**

There are levels of consciousness beyond the material plane at the spiritual level of Neshamah. These are the levels of consciousness in which Adam and Eve resided before their descent into the material plane while they were still behind the curtain. They had bodies of light that protected their sparks of life while they lived in that level of consciousness. At that level of consciousness they had possession of Eternal Life. But then Adam and Eve began to turn their attention to fulfilling their desires to experience the world of sensuality. As their attention turned to visualizing greater interaction with the material world, and all of the mysteries of nature, they began to experience the darker sides of moods and emotions. As a result their bodies of light became extremely dim as they slipped ever deeper into material consciousness, and emanated a body of solid material flesh.

With a body of material flesh came death. Why?? Because the material world is a world of decay and death. There is a definite correlation between the amount of time that a human mind dwells on material, fleshly, sensual thoughts and the depth or level of density that encases that same body-mind-spirit. There are people on this earth today that are so debased, demented and depraved that no light at all emanates from their spirit. For all practical purposes they are the living dead. There is a Blackfoot Indian

Proverb that states; **"life is not separate from death. It only looks that way."**

Adam and Eve were warned that if they pursued the fruit of the knowledge of Good and Evil they would die. Their Adversary told them that they would not die. What actually happened was that they became so entwined with their material world of sensuality that they identified themselves with their own material bodies. Ruach and Nephesh became dominant in their lives. They developed a consciousness of death in which they realized that their consciousness of experience and memory would eventually come to an end. In reality, when Nephesh ceases to animate the body, the body immediately begins to disintegrate by a process called Death. All of the experiences of the mind is withdrawn as the Spark of Life exits the body with Neshamah and Ruach into the spiritual realm. Neshamah (the ideal self) and Ruach (the false self) are then compared to see what was accomplished in life and what is still lacking. And from this information a new mission into the material realm is developed and launched. This is called Reincarnation.

In our space shuttle analogy, material life begins as the space shuttle blasts off and leaves the earth with a mission to fulfill. When the mission has been completed the space shuttle leaves the realm of space and begins its descent back into the atmosphere of the earth. This would correspond to the experience of death. But the results of the mission would return to the earth for further analysis and the planning of future missions. The ultimate goal of the missions into space would be to have a permanent presence in space, thus reducing the need for the temporary flights of the space shuttle. And so it is with our physical experience in the material realm.

We live in bodies of temporary material flesh, which limit the light of our higher bodies. But this can be remedied. By spending more energy in the mental and higher planes of consciousness we can increase the vibration rate of our material bodies and they will become less dense. This is because we can increase the intensity of the body of light within us. By advancing closer to Divine Thought we can begin the change that *transforms* us to become more like Divine Thought, both in power and in knowledge and wisdom. As our levels of consciousness expand the body of light grows ever brighter and stronger. As a result this body of light can have beneficial effects even on the state of health of the physical material body.

Paul summed it up in I Corinthians 15:49-51, **"As we have borne the image of the earthy, we shall also bear the image of the heavenly. Now this I say brothers (and sisters), that flesh and blood cannot inherit the kingdom of God; neither does corruption (decay) inherit incorruption (unending existence). Behold, I show you a mystery; we shall not all sleep but we shall all be changed . . . "**

The simple truth is that the material body of flesh and blood cannot live forever because it is a body of death. It is only the *body of light* that can remove you out of the wheels of Reincarnation. When the material body has been transformed to a level of energy sufficient to permit it to exit the domain of the material realm without destruction and death, then Reincarnation will be no longer necessary.

The Gospel of Phillip wrote it this way: **"Those who say they will die first and then rise up are in error. If they do not *first* receive the Resurrection while they live, when they die they will receive nothing."**

The Tree of Life is a stairway, or a pathway from the material plane of consciousness to the higher planes. There are many planes of spirituality and levels of consciousness. Each of these is an energy-information level beyond the human plane of consciousness. God, or Divine Thought is at the highest level of consciousness. The human soul was designed to span each and every plane. By using the Tree of Life you can gain access to a variety of levels of knowledge and a variety of powerful transforming energies. By focusing these energies you can change your life and even . . .

Transform your own environment.

Soul is the Bridge between GOD and Man

1. The Endless World is the source of our existence and is where we came from as Sparks of Life.
2. Sparks of Life posses a "desire to receive" and a "desire to share." This defines their purpose.
3. A material world was prepared to provide an opportunity for Sparks of Life to pursue their purpose.
4. Soul is a body of protection and communication for Sparks of Life while in the material world.
5. There are 3 elements of the soul to protect the Sparks of Life in 3 different worlds.
6. *Neshamah*, the highest manifestation of soul, has access to the knowledge and wisdom of "God."
7. *Ruach* is the manifestation of soul that gives us our individual personality.
8. *Nephesh* is the manifestation of soul that is often called "human nature."
9. Each soul has an opposite-sex soulmate that becomes the object of a life-long search.
10. Each soul can have 2 bodies; a material body and a spiritual body of light.
11. Death is the disintegration of the material body when the body of light is removed.
12. Eternal Life is achieved when the body of light is powerful enough to transform the material body into Light.
13. The Tree of Life is the source of knowledge and transforming energies.

Chapter 5

The Tree of Life

When Abraham discovered the knowledge of how the Universe came into being, he put away his idols and relics of religious worship and moved into a whole new world of understanding. Abraham grew up in the kingdom of Chaldea, at the port city of Ur. They called their gods and goddesses ELOHIM, which simply meant Deities. Chaldeans believed in Astrology and worshipped the stars, which they believed influenced their lives and destiny. All of the surrounding nations also had Elohim that they believed were their protectors and influenced their destinies.

The Hebrew people were no exception. They took Abraham's discovery to mean that there was one Supreme Elohim who ruled over all of the others, and it was He that they worshipped. A voice told Moses that His name was *I AM THAT I AM,* which has come to be known as YAHWEH, or in English, Jehovah. But Abraham saw a much broader picture. Abraham discovered that there was a pattern for everything that exists, has existed, or ever will exist. This pattern exists in the form of thoughts. Therefore, Abraham realized that God [Elohim] does not dwell in the Universe, but IS the Universe. In other words, without the "thought patterns" the Universe would cease to exist. All of the forces and laws of nature and of the Universe would cease to exist. Therefore, Abraham must have concluded, The Supreme Being must be pure thought. That is when Abraham began to refer to the Supreme Being with a meaning that we will call *DIVINE THOUGHT*.

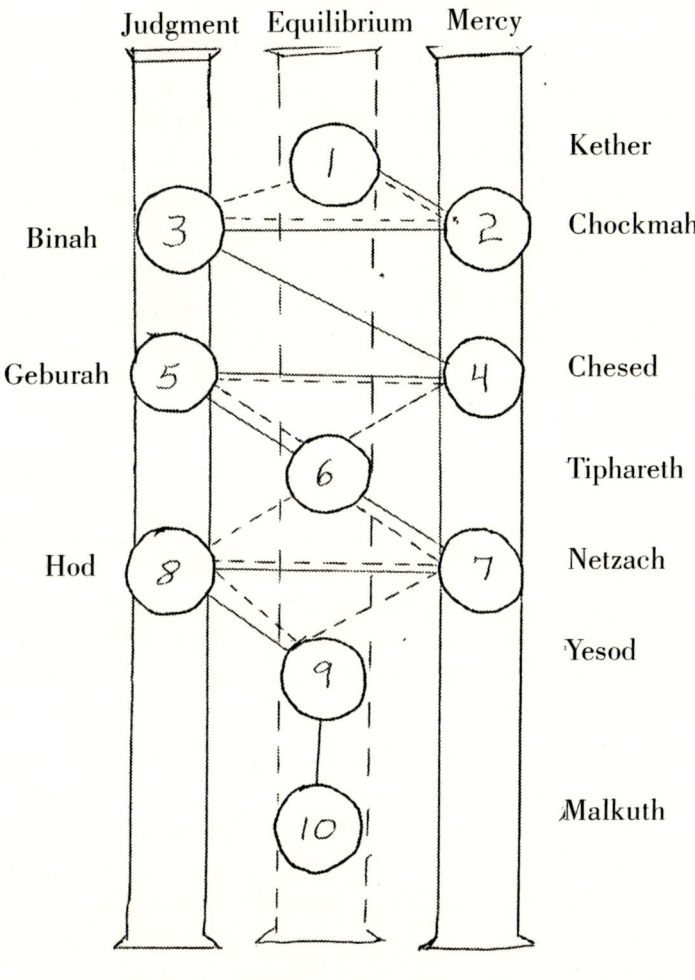

The 10 Spheres

DIVINE THOUGHT manifested itself through 9 stages before it reached the 10th and final stage of physical manifestation. Each of the 9 stages successively acquired greater density as they added definition to the thought patterns, or thought forms being manifested. Each stage is a sphere of influence of a specific classification of energies that give substance and definition to the thought form being manifested. These 10 Spheres are called The 10 Sephiroth. The Tree of Life is a symbolic glyph that classifies these 10 Spheres of energies and organizes them into an understandable formula that is useful to anyone who wants to use them. Since the Tree of Life exists within the human form, it can be accessed by anyone. This is "the God Within." The glyph of the Tree of Life shows the stages of manifestation as 10 spheres, or circles organized along what is called the Path of the Flaming Sword.

The 10 Sephiroth are arranged along 3 vertical parallel columns or pillars. The right pillar is called the *Pillar of Mercy* and defines the active male potency. The left pillar is called the *Pillar of Judgment* and it defines the female passive principle. The central pillar is called the *Pillar of Equilibrium*, and its function is to balance and harmonize the Pillars of Mercy and Judgment.

The first Sphere is at the top of the central pillar, and it is called Kether. The second Sphere is 30 degrees from vertical below Kether on the right pillar. It is called Chokmah. The third Sphere is horizontally across from Chokmah on the left pillar and is called Binah. These three spheres form an equilateral triangle with Kether at the top.

A second mirror image triangle is formed below the first one, with spheres 4, 5, and 6. The fourth sphere is on the right pillar and is called Chesed. The fifth sphere is horizontally across from Chesed on the left pillar and is called Geburah. The sixth sphere is below on the central pillar, thus forming an equilateral triangle pointing downward. The sixth sphere is called Tiphareth.

Below Tiphareth are spheres 7 and 8. The seventh sphere is on the right pillar and is called Netzach. Horizontally across from

Netzach on the left pillar is the eighth sphere called Hod. Spheres 6, 7, and 8 form a third triangle with Tiphareth at the top. The ninth Sphere is below Netzach and Hod on the central pillar and forms a fourth triangle, pointing downward. Sphere 9 is called Yesod. Directly below Yesod, also on the central pillar, is the tenth sphere called Malkuth.

Looking at the 10 Sephiroth on the 3 pillars you will see that the right pillar contains the 3 Sephiroth of Chockmah, Chesed, and Netzach. The left pillar contains the 3 Sephiroth of Binah, Geburah, and Hod. The central balancing pillar contains the 4 Sephiroth of Kether, Tiphareth, Yesod and Malkuth. This is a picture that represents the Tree of Life.

All of the forces and energies of the Tree of Life condense down into the dual principles of Mercy and Judgment. The right pillar is the Pillar of Mercy, which can bestow unlimited mercy. Unlimited mercy gives total freedom to do anything, no matter how heinous or violent, without any hint of consequences, or restraint. The left pillar, the Pillar of Judgment, can bestow unlimited restraint. Unlimited judgment brings total confinement or restraint, like a straitjacket, no matter what you do to try to change the circumstances.

The central pillar, the Pillar of Equilibrium, is the mediator between the two extremes of mercy and judgment. The whole purpose of the Tree of Life is to provide balance between these two extremes. Many people believe that the "God of the Old Testament" was a God of rules and laws, but this was not the case. In reality, the people of Israel could not exist very long in a state of unlimited mercy. Since they so quickly became self-destructive, it was necessary to impose the restraints of judgment upon them. The state of unlimited mercy lasted just one year in the desert before the rules of judgment were enumerated at Mount Sinai. But the *desires of God* were very clear. He said in Hosea 6:6, **"For I desired *mercy* and not sacrifice; and knowledge of God more than burnt offerings."**

And in Hosea 12:6, **"Therefore, turn to your God; keep *mercy* and *judgment*, and wait upon your God continually."**

YAHSHUA, the Son of God, chided the Pharisees of his day on this very point when he said in Matthew 23:23, **"Woe unto you, scribes and Pharisees, hypocrites! For you pay tithe of mint and anise and cummin, and you have omitted the *weightier* matters of the law, judgment, mercy and faith. These you should ought to have done and not left the other undone."**

Do you see the train of thought here? *YAHWEH*, in the Old Testament, said that it was more important for the people to show *mercy* to one another than to offer burnt offerings. In fact He wanted them to keep (or practice) mercy and judgment continually. YAHSHUA told the scribes and Pharisees the very same thing! He said that the balance of judgment and mercy through faith was much more important than upholding the letter of the law. You see, judgment, mercy and faith are the very foundation of the literal laws, ordinances and statutes.

Remember when asked which was the most important of the laws, YAHSHUA said, you shall love God with all your heart, your soul and your mind and love your neighbor as yourself? To do so requires a balance of judgment and mercy. Those who do not practice mercy will reap the same consequences on a cosmic level, as James 2:12-13 explains, **"So speak ye, and so do, as they that shall be judged by the law of liberty (tree of life). For he shall have judgment without mercy who has shown no mercy; and mercy rejoices against judgment."**

Are you beginning to see how the human soul is so closely tied to the Universe? The Universe emanated from the Tree of Life. The foundation of the Universe is the multitude of laws and forces that interact to keep it in perpetual balance and equilibrium. At the same time, the pillars of the Tree of Life are judgment and mercy. These are laws and forces that keep the Universe, and the human soul in balance and equilibrium.

In the Stars Wars movie series, the Tree of Life is called The Force. It has a Light and a Dark side, and it also has feeling, as when Obi Wan said, "I feel a disturbance in the Force, as if millions have cried out in pain." The *"will* of The Force" has also been mentioned throughout the series, which indicates that The

Force has intelligence and purpose. And the salutation, "May The Force be with you" expressed the hope that you would find favor with the source of that Intelligence. So the next time you watch a Star Wars movie, try to discover the many qualities that are attributed to The Force, and realize that this is just one more way of describing the Tree of Life!

Carolyn Myss, Ph. D. in her book *Anatomy of the Spirit* describes a conversation she had with Sogyal Rinpoche, author of *The Tibetan book of Living and Dying.* He pointed out to her that "when any spirit (soul) leaves the earth the entire energy field is influenced. And when a very powerful spirit leaves, the influence upon the earth is even more dramatic." The Star Wars incident described by Obi Wan illustrated that when a whole planet of people died, it created an immense reverberation throughout the entire Universe of fear, panic, dispair and pain. And these emotions were strongly felt by Obi Wan Kenobi, who was sensitive to the emotions that were flowing through the Force.

Candace B. Pert, Ph. D. is a biochemist who produced a cassette taped series in 1997, called *Molecules of Emotion.* She helped to pioneer the discoveries demonstrating that "Biochemicals are the physiological substrates of emotion." She goes on to explain that every emotion in the human body is produced by a precise chemical that bonds to specific cells in the body that are receptive to that emotion. The emotions they have studied are any and all feelings, sensations, thoughts, drives and even altered states of mind. Drive states are emotions like hunger, thirst, urge to defecate and urinate, and sex. Even spiritual manifestations have been shown to be biochemically controlled, such as awe, bliss, serenity and inspiration!

According to her research, here is how the human body works. Every cell in the body has a number of receptors on their surfaces that are like tiny keyholes. The number of receptors varies on each cell but all cells have at least one or more of them. They have identified at least 72 different receptors, and each receptor is responsible for a different and specific emotion or response. These receptors will determine which thoughts, for example, will be activated in the conscious mind, and which will remain under the surface as unconscious thought patterns.

There is a tiny key, about the size of a virus, which fits and unlocks these receptors. Each of the 72 keys is differently shaped so that it fits ONLY its own keyhole or receptor. These keys, which are called *peptides*, are manufactured somewhere in the body when they are needed to trigger an emotional response. Receptors are by far, most numerous in the areas of the brain that control the emotions. Peptides are actually couriers that carry *information* to the body cells. Receptors are designed to receive specific kinds of information. When a peptide unlocks a receptor, then the appropriate information is transmitted to that cell. A constant exchange, processing and storage of information is the stimulus that animates the body. Peptides are messenger molecules that link the brain, the body and all human behaviors. Every cell in your body has the capacity to store useful information for later processing or exchange with other cells.

You may recall from chapter one, Dr. Pearsall's discovery that a transplanted heart brings with it information and memories from the heart donor. Most heart (and other organ) recipients have an awareness of that information from the donor. It is the various peptides still contained in the heart that unlock the cells and carry the information from the donor heart to the recipient brain. For awhile the recipient feels the emotions, feelings, sensations, thoughts and memories of the organ donor. Eventually these emotions become integrated with those of the recipient. You see, Emotions are the result of millions of exchanges of information taking place at the cellular level every second.

Dr. Candace Pert asserts that the human mind is a flow of information moving among the cells, organs and other body systems. The mind, therefore, is not limited to the brain but occupies the entire body structure. In other words, the mind becomes the body. The body is the outward manifestation of the mind into physical space. It is the emotions that connect the physical matter and the mind. The mind can influence the physical matter of your body, and your physical body can influence the operation of your mind. It is the emotions (flow of information) that provide this 2-way interaction of mind and matter.

Dr. Pert says, **"We are a dynamic system with a constant potential for change, in which self-healing is the norm and not a miracle."**

Your Ruach or Ego is the mind, which exercises control of the free flow of information at the cellular level. Every action that the body makes, whether voluntary or involuntary, is based on an exchange of information at the cellular level. Every thought or emotion in your body or mind is caused by information carried by biochemical peptides to or from the affected cells of the body.

Dr. Pert goes one step further and says, **"We are an integrated body and mind with emotional intelligence."** The flow of activating information, or emotions, is like a subtle electrical current flowing through the body. This has long been recognized by Eastern Religions as Chi, by Wilhelm Reich as Orgone Energy, and the Kabbalah as subtle energy. This flow of energy, which keeps the body systems operating at optimum levels of energy and health, has intelligence. Now we can understand why. It is because this current is a flow of information with a specific purpose that is controlled by the body, brain, or the mind. And this brings us back to the Tree of Life.

The flow of energy from the Endless World of Divine Thought first produced the Adam Kadmon, and from there it produced the Four Worlds or Planes of Consciousness. In the previous chapter we explored the Four Worlds as they relate to the human soul. It is the soul that infuses a lifeless body with life, vitality and animation. All of life is sustained by a continuous flow of energy-information. And it is the Tree of Life through which all of that energy-information flows.

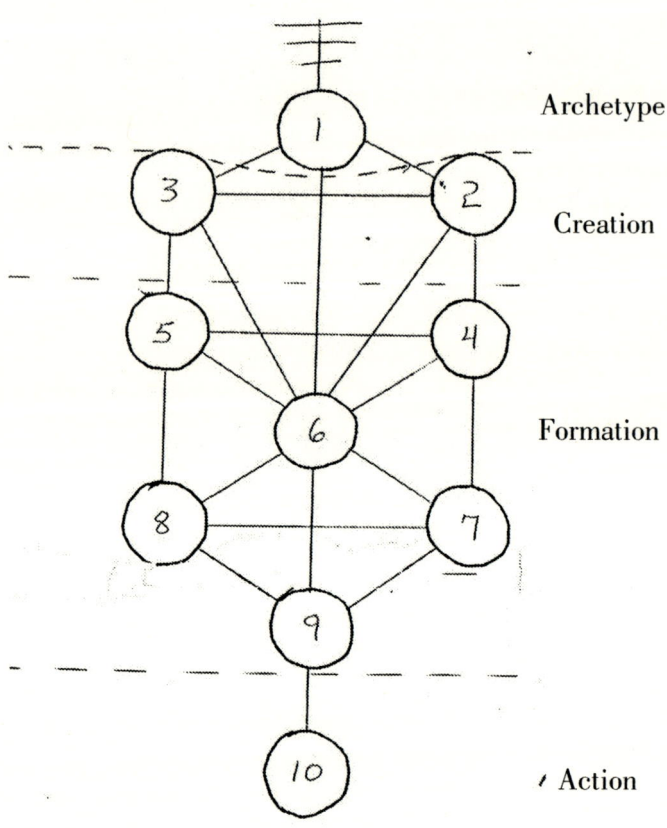

Archetype

Creation

Formation

Action

tree of life and the four worlds

The first of the Four Worlds is *The World of Archetypes*. It is a world of very high intensity energy that is the source of all of the information that the Universe could ever retain. The first Sephiroth of the Tree of Life, Kether, contains this world and is the source of all information that will ever be available in the Universe. Kether means Crown, the source of everything.

The second world is *The World of Creation*. This is the

mental plane where the raw information is organized into specific thought forms. These thought forms can be accessed by the human mind and used. This is still a world of high intensity energy but it is the energy of higher levels of thought. Two Sephiroth control the flow of energy in this world. They are Chockmah (Wisdom) and Binah (Understanding). Without Wisdom and Understanding the flow of information would be useless. The understanding of information determines how it should or could be used. Wisdom determines when and why it should be used. Now you know why the sustaining flow of energy of and within the Universe is called intelligent energy. Any intelligence we humans have is derived from that *universal intelligence!*

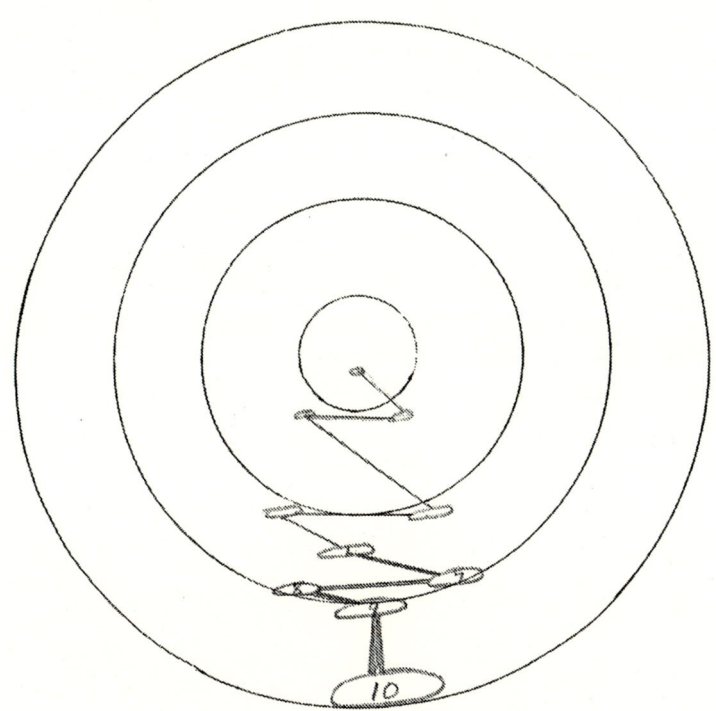

the four worlds and the tree of life

The third world is *The World of Formation*. This is the world in which our thoughts and our level of "intelligence" reside. This is the part of the Tree of Life that contributes to our quality of life, our character and our moral fiber. This is the part of the Tree of Life that determines who we *think* we are! The next 6 Sephiroth make up this world of flowing energy-information. The first 2 of these 6 are Chesed (Mercy) and Geburah (severity, or judgment). If you think about it, most of our human endeavors center on learning how to balance mercy and judgment. By reaching to the higher levels of comprehension for wisdom and understanding, we can more easily achieve this goal. The third of these 6 Sephiroth is Tiphareth (Beauty). It is centrally located between Chesed and Geburah above it, and Netsach and Hod below it. Beauty is achieved by balancing the energies of those four Sephiroth. Netsach (Victory) and Hod (Glory) are also energies that need to be balanced against one another. The lowest Sephira in the World of Formation is Yesod (Foundation). When all of the above mentioned Sephiroth are in balance, then you have a very solid foundation on which to base your life. This is the purpose of the Tree of Life.

The fourth world is *The World of Action*. This is the material world. Its Sephira is called Malkuth (Kingdom). You can have the best of intentions in the world, and think the purest and most noble of thoughts, but if they are not put into action, they are useless. All of the 10 Sephiroth are connected to each other through connecting pathways. The pathways that connect the 10 Sephiroth from Kether, zigzagging down to Malkuth, are called the *Path of the Flaming Sword*.

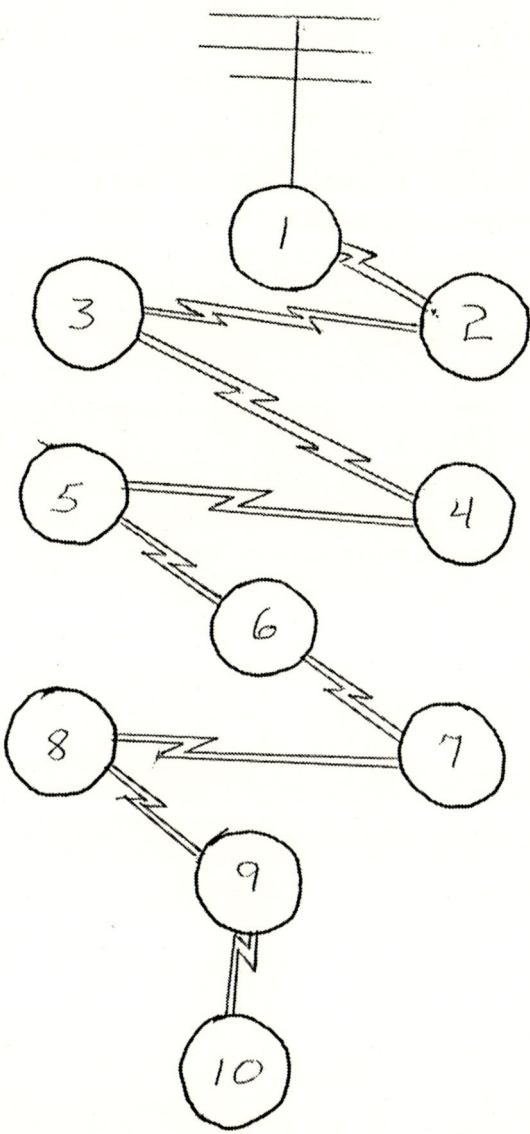

Path of the Flaming Sword

You can begin to activate the Tree of Life within your own Consciousness by using an exercise called "The Path of the Flaming Sword." This can be done by *visualizing* your body superimposed upon or within the 10 Sephiroth of the Tree of Life. You should see Kether at the top of your head with the rest of Sephiroth dispersed down throughout the body with Malkuth at the bottom of your feet.

Visualize a brilliant ball of light beginning at Kether and moving one Sephira at a time until it reaches Malkuth. As you move from one Sephira to the next visualize a brilliant blue flame, like lightning, burning a path to the next sphere, or consciousness level, until you reach Malkuth. As you continue to practice this exercise it will become easier and the energy flow will become stronger. Once you see the 10 Sephiroth all aglow, then visualize the energy expanding to envelop your entire body in pure white light. This exercise will begin to energize your body and make available some of the energies you need from each level of Consciousness. This is a good exercise to develop greater mental strength and assertiveness.

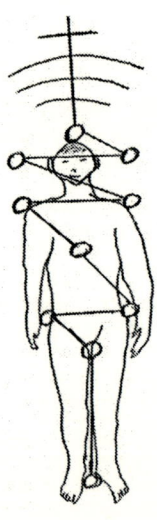

tree of life within

In the following chapters you will learn some of the energies and qualities that are available at each level of consciousness. You can use this knowledge to better concentrate on attaining the attributes and qualities that you need or desire to have. You are now beginning to use the tools of . . .

Meditation and visualization.

The Tree of Life

1. The Tree of Life is a symbol of 10 successive reductions in highly intensive energy vibrations.
2. These are 10 spheres or Sephiroth that categorize qualities of energy-information.
3. The 10 Sephiroth are arranged along 3 Pillars; which are judgment, mercy, and equilibrium.
4. Dr. Candace Pert discovered that emotions and thoughts are bio-chemical reactions that transfer cellular information.
5. This flow of energy-information is called subtle energy, which is the foundation of material life.
6. You can learn how to use this *intelligent energy-information* for your own education, healing, and spiritual advancement.
7. The Tree of Life is the conduit through which all energy-information and intelligence flows.
8. "The Path of the Flaming Sword" is an exercise to use to activate the Tree of Life within.

Chapter 6

MALKUTH #10

Our studies of the Principles of Kabbalah have shown that the Universe has emanated from higher levels of energy down to levels low enough to manifest into material form. Furthermore, this energy has been described as being perfect intelligence. The most ancient of mystical religions, including the Kabbalah and the Bible, have taught this idea. We refer to this intelligent energy as Divine Thought.

The Kabbalah also teaches that human *souls* emanated simultaneously with the Universe. The soul contains elements that allow it to communicate with Divine Thought, or Universal Intelligence that flows throughout and permeates the entire Universe. We also have learned that the ability to communicate with Divine Thought was lost or became much more difficult as the Living Souls descended ever deeper into the material realm of sensuality and baser emotions. The soul actually became imprisoned within the material realm. But the good news is that it is possible to begin to reclaim that lost ability of communication with Divine Intelligence.

To begin to restore our lost art of communication with Divine Intelligence, we should begin right where we are in the material realm or the World of Action. Malkuth is the Sephira that has been assigned the number 10. It is the bottom Sephira of the symbolic Tree of Life. This is the sphere of sensuality and sensation. It is the only Sephira that does not form a direct triad with any 2 other Sephiroth.

Malkuth is a container for the Emanations of all of the other 9 Sephiroth. The Tree of Life is somewhat like a quality separator. As the pure energy of Divine Thought enters the highest Sephira of the tree of Life, the purest qualities are filtered out and remain at that level. Everything else descends to the next level, and to the next, where other qualities of less purity are filtered out and remain. Any qualities not retained by the 9th level fall into the level of Malkuth. For that reason Malkuth contains the densest and coarsest of human qualities. But that is the nature of the material realm.

Malkuth is anchored at the bottom of the Pillar of Equilibrium, the Central Pillar. This explains why the laws and forces of nature always seek a balance or equilibrium. You might wonder why Man, who lives within the balance of nature, is not a part of that balance. This is because Nature has instincts that are programmed to seek balance, whereas Man has an intellect that he can use to override his instincts. The instincts of Man are called human nature. Intellect is an element of the soul that animals do not have because they do not possess a Tree of Live within. Because intellect gives Man free will to use his human nature against his fellow man, he can unwittingly destroy the balance of nature in the process.

Most books that teach you about the Tree of Life teach you only about the positive virtues of the Tree. They gloss over the negative aspects of the Tree. But as you now know the Universe is a system of duality. And you also know that for every positive there is a negative. So remember, whenever you are presented with something new, realize that there is always an equal opposite to it. George Lukas did an excellent job of demonstrating this fact in *Star Wars* with the use of "The Force." The Force is just another word for Tree of Life. It has a Light side and a Dark side.

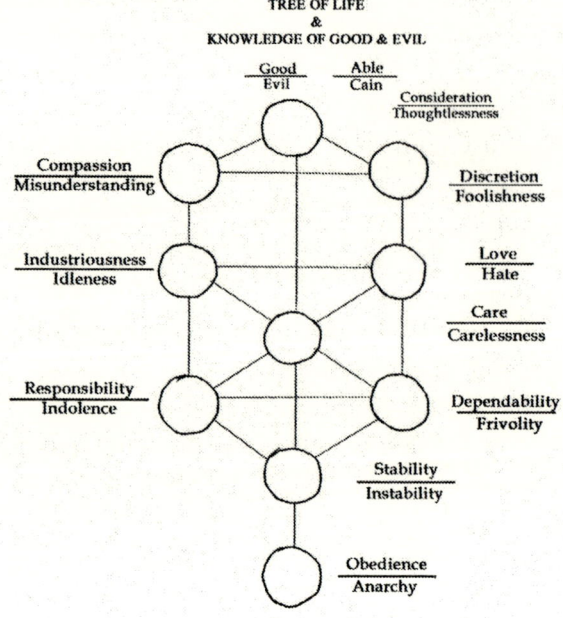

TREE OF LIFE
&
KNOWLEDGE OF GOOD & EVIL

Abel and Cain tree

Obi Wan Kenobi Taught Luke Skywalker how to use the Light side of the Force. Both he and Yoda were very clear that the Dark side of the Force would be easy to slip into if Luke did not have a clear mind and Conscience. It was called being "seduced by the Dark side." When the Emperor took on the task of training Luke in the Dark side of the Force, he was inflaming the passions of human nature, which is the seat of the Dark side.

The Kabbalah explains that the Dark side of the Tree of Life is called the Tree of Knowledge of Good and Evil. This is the area of human nature. It includes, but is not limited to, the emotions of anger, rage, jealousy, resentment, malice, filthy communication, lying, coveting, lust, fear and dispair. If someone has a problem with emotions from the dark side, they can use the Tree of Life to help gain control of those emotions.

Malkuth is a level of consciousness that is closest to our natural waking consciousness. The main *virtue* of Malkuth is *discrimination*. This is where you will work to gain better discrimination in your thinking. You can learn how to better discern between good and evil instead of relying on someone else's interpretation. Discrimination can help you to discover what is best for you in your life decisions. At this level of consciousness you can even learn more about your own health problems and how to handle them. This is the level at which all illnesses manifest in the material plane, and this is where you would begin to search for the cause of those illnesses. In other words, greater discrimination can lead you to even greater self-discovery.

The *vice* at this level of consciousness is *avarice* and *inertia*. Avarice is a greedy desire for gain. It is a strong desire to gain money and/or property by any means necessary. Inertia is the tendency to remain in the present state of mind or being. It is a resistance to beginning a new project or even beginning to make changes in ones life or environment. Some might call it a state of laziness. If avarice or inertia is a problem for you then this is the level of consciousness you would use to combat that problem.

Malkuth is the world of the 4 elements of matter (earth, water, wind and fire) and this is where you would work to bring balance to those elements. This is also the world of the 5 senses and everything that can be perceived by those 5 senses. The 5 senses relate to the material world, which is the Consciousness level of Malkuth.

Malkuth is the level of consciousness of the tribal belief system. Your first memories are those of being a loyal member of a tribe or group and you adhere to all of the beliefs of that tribe. At this level you fully believe that your tribe is infallible in its beliefs, teachings and laws. People at this level of consciousness are usually the first to fall victim to contagious diseases that spread throughout the tribe because of the shared belief that all will share in that disease. The shared belief system rests at a subconscious level in which all of the members are anchored. Today, a tribe could be a family, a community, a religion, or even a government.

Malkuth is the level of consciousness at which the *Holy Spirit* manifests. Kabbalah teaches that the Shekinah is the executive energy of Binah in Malkuth. This means that the feminine energy of Divine Thought must manifest itself at the level of consciousness attainable by the mind of Man. Therefore, the Shekinah energy usually manifests itself in a tribal or group setting as one of the many powers or *gifts* of the Holy Spirit. Some of the gifts of the Holy Spirit are explained in I Corinthians 12:1-12. The expressed purpose of manifesting one or more of these spiritual gifts is to aid the person, and their tribe or group, in spiritual advancement. Pentecostal Churches are usually associated with manifestations of these gifts, because their religious services are designed to increase the flow of the energy of Malkuth among the group's participants. Those who successfully tap into that increased level of energy will manifest that energy in one of several forms, which the Bible calls the gifts of the Holy Spirit.

One gift, or form of this energy, is wisdom, and another is the word of knowledge. One who manifests wisdom usually displays wisdom far beyond his normal capacity. It is the same with the word of knowledge. The person will usually display specific knowledge that he/she never possessed before. That knowledge just pops into his/her head out of thin air, to be useful for someone. Another gift of the Holy Spirit is faith. Faith is described in the Bible as *"evidence of things not seen, and the substance of things hoped for."* In a court of law, evidence is often presented to substantiate a charge against someone about an event that was not seen by anyone. The jury either accepts, or rejects that evidence based on faith that the witness is either reliable or not reliable. A person manifesting the gift of faith just knows whether or not something is true without any outside evidence. It is akin to the gift of the word of knowledge. Other energy manifestations of the Holy Spirit are healing, miracles, and prophecy. In the healing manifestation, a person strongly desires to be healed of a malady. The Healer is able to transmit a strong charge of energy, usually through touch, which is used by the subconscious mind to effect the healing process in the body. Miracles are usually obtained through a similar process. Prophecies are generally obtained for the group by someone who sees mental images of a future event.

Discernment of spirits, speaking in tongues, and interpreting tongues are the other manifestations of the Holy Spirit that are mentioned in I Corinthians 12. Discernment of spirits basically means to evaluate the intentions of a spirit whether for good or ill. Actions are then taken based on that discernment. Speaking in tongues is often speaking in a language that is foreign to the speaker. Interpreting tongues is interpreting the words from a foreign language by someone who is unfamiliar with that language.

These manifestations of Shekinah energy are all derived from one source of energy that is available at the Malkuth level of consciousness. The manifestations are usually based on group participation, but that is not always the case. Sometimes a person can use his or her gift at will.

Each of the 10 Sephiroth has a name of God attached to it that describes the attributes of that Sephira. Divine Thought cannot interact with the material world without destroying it. Therefore, provisions were made for Divine Thought to be able to manifest some of its energy and presence at each level of consciousness.

The name of God is the most powerful aspect of each Sephira and is the most spiritual manifestation of that level of Consciousness. The name of God at Malkuth is *Adonai Ha Aretz*. It can be used as a powerful mantra at this level. This is the God of the kingdom of Man, which influences our affairs on the material level. It means *"Lord of the Earth and the Visible Universe."*

There are also classes of angels assigned to work in each of these 10 levels of consciousness. The class of angels in Malkuth is called Ashim. There is also an Archangel in charge of each level and if you meet a Being while you are working in a particular level of consciousness, it might be him. The *Archangel* for Malkuth is named *Sandalphon*. He is called the Approacher and the Prince of Prayer. This is also the level of elemental spirits, which have given rise to the belief in fairies, elves, gnomes and leprechauns. You might also encounter them in your work at this level of consciousness. They are not to be feared, but neither are they to be trusted.

To access the energies of the Tree of Life at this level you may visualize the Path of the Flaming Sword as before. Determine which attributes of Malkuth you wish to awaken and then feel the energy as it flows down the tree to Malkuth. Feel the sensation as the energy expands to encompass your entire body. As you absorb the energy, just *know* that the attributes that you desire are being awakened. The energy is following the pathway that is created by the thought of your strong desire to receive. As you begin to exercise these awakening attributes they will become stronger and will begin to effect the desired changes in your life. We came into the material world to make changes in ourselves, and this is the level at which we can begin to . . .

initiate those changes.

Malkuth # 10

1. We can learn how to communicate with Divine Thought to begin to change ourselves.
2. Malkuth is the level of consciousness of Human Nature.
3. The *virtue* of Malkuth is *discrimination* or discernment.

4. The *vice* of Malkuth is *avarice* or *inertia*.
5. The Dark Side of the Tree of Life is called the Tree of Knowledge of Good and Evil.
6. Malkuth is the consciousness of "the Tribe" or the infallibility of group belief.
7. The Holy Ghost (Spirit) manifests its presence at the level of Malkuth.
8. The Tree of Life can be used to expand the attributes we want in our lives.

Chapter 7

YESOD #9

Yesod is anchored just above Malkuth on the Central Pillar of the Tree of Life. It is the level of consciousness through which virtually all energy from above flows. It is at this point that everything for the material level is organized, filtered and arranged for man's use. This is also the seat of human intuition. This is the level of consciousness of the Moon, which is the hidden side of conscious thought. Many of the unknowns that influence our emotions and feelings are hidden here quietly influencing our conscious thoughts. All of the rhythms, cycles, and fluctuations that sustain life originate here. And this is where you can begin to learn more about the mysteries of organic growth and life.

Yesod is the level of the astral plane and the etheric plane. The Consciousness level at Malkuth is influenced by microscopic chemicals and the information that they carry, whereas in Yesod it is composed of one fluid substance. Everyone, whether animal or human has an etheric field that can be observed by some sensitive people. In some people the etheric field extends about one-eighth of an inch outside of the body that gives it a glowing appearance. In very healthy and energetic people the etheric field can extend beyond the body up to 6 inches or more. When a person dies it can take up to 72 hours for the etheric field to completely and fully dissipate.

The etheric field is actually the essence of the body's aura. The human body has an aura, which is actually an extension of the soul. The aura can extend out from the body from 6 inches to several feet. It is composed of the colors of the rainbow and these are indicators of the body's health, mental capacity, character, and present attitude. The aura is with you throughout your life and is an indication of your spiritual status or present level of attainment in this life. This information is all stored in your mind as mental images.

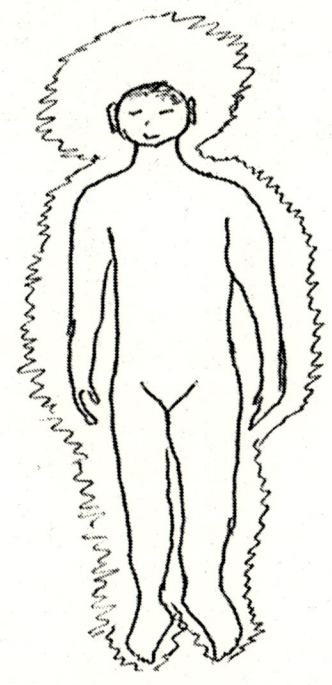

human aura energy field

Everything you know and think is stored in your memory as images. Even the words you speak are converted to images in your mind. Thought forms are a collection of images tightly bound to a common theme or idea. It is at this level of consciousness where these images are built. You have a conscious or subconscious *link* to every person you know. That link is based on the images you have seen, felt and thought toward that person. Even feelings and emotions are stored as images. *Thought is energy.* Thought forms are energy that is molded to create a specific function, and they are sustained by a constant flow of incoming energy. Each link you have to a person, pet, idea, emotion or feeling sustains a continuous flow of energy. Under many conditions that flow of energy can be damaging to your mental and physical health if it becomes a mental drain upon your personal energy supply. This can become a source of emotional and even physical illness.

In the area of personal development, your subconscious psychic and biological functions can be awakened and stimulated in this Sephira to achieve greater activity. Thought forms can be created by your own mind to attract and increase a greater flow of energy to your mind and body from higher levels of consciousness. You can begin to learn to acquire more energy and to achieve more control over your life at this level.

Yesod is the universal reproductive power in nature. This is where creativity and reproduction begin to manifest. The Kabbalah says, **"Everything shall return to its foundation from which it has proceeded. All marrow, seed and energy are gathered in this place. Hence, all of the potentialities which exist go out through this."**

Have you ever planted seeds in a garden or a flowerpot? You plant the seed in the ground and a few days later, a new plant rises up out of the earth. You know that the seed "dies" and then new life sprouts from that seed. Where does that life come from? How does the tiny seed produce a new green plant? The seed contains within it an image of the plant. The microscopic image contains all of the information that is needed to complete the entire life cycle of the plant and seed. The information determines the size, color, texture, timing parameters, and an infinity of other defining characteristics. Every cell of the plant and seed contains the image of all of this information. This information can be accessed through the DNA of the plant. The DNA information is all stored as an energy image in the level of Yesod. All of the potential energy of the growing plant emanates from the etheric energy level of Yesod, and when the plant dies and decays, the energy returns to its origin in Yesod.

The growing science of genetics is increasing its ability to manipulate genetic information to effect changes in plant growth and other hereditary factors. The human body has all of the complexities of its existence stored in the genetics of its DNA as well. The Human Genome Project is focused on accessing the genetic information stored in the human DNA, and learning how to manipulate that information to create hereditary changes within the human body. Most of this research would ostensibly be used to cure illnesses. However, A sharply focused mind can accomplish the same curative results in the level of Yesod.

All *emotional* based illnesses manifest from this level of consciousness. Emotional based illnesses manifest from strongly held image thought patterns that create restrictions of energy flow in specific parts of the body. The restricted energy flow severely curtails the flow of nutrients and other vitals to the cells of the body and they eventually become polluted and die. This allows consumptive diseases like cancer to develop and spread from the restricted part of the body.

Louise Hay, writes in her book, *Heal Your Body*, **"Both the good in our lives and the dis-ease are the results of mental thought patterns which form our experiences. We all have many thought patterns that produce good, positive experiences, and these we enjoy. It is the negative thought patterns that produce uncomfortable, unrewarding experiences with which we are concerned. It is our desire to change our dis-ease in life into perfect health . . . . Our consistent thinking patterns create our** experiences. Therefore, by changing our thinking patterns we can change our experiences."

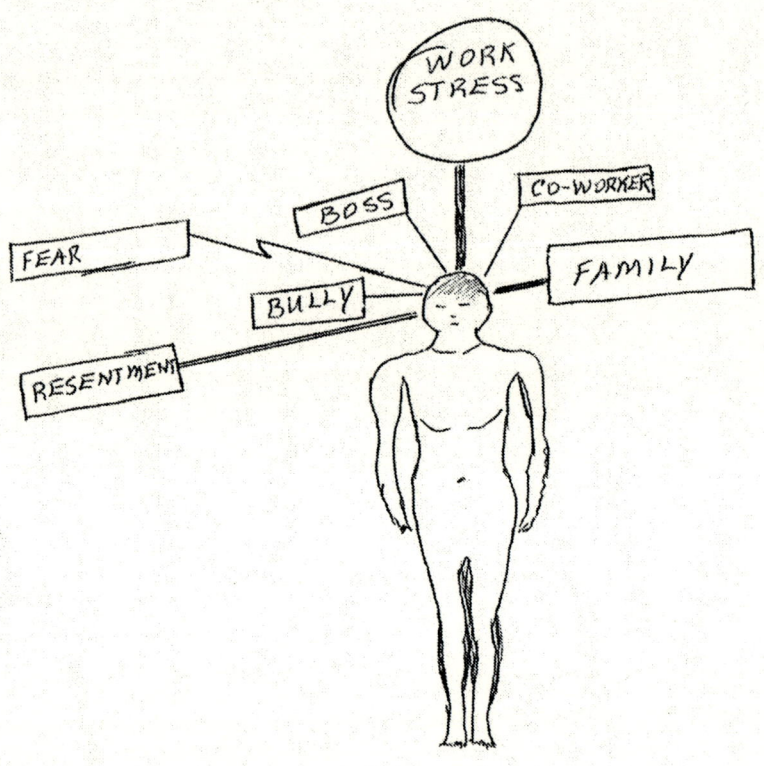

Emotional links to thought forms

Is your life filled with good experiences or bad? As Louise Hay points out, all experiences come from your general day to day thinking patterns. Your emotional health patterns are manifested from your day to day experiences. You have a link to each person in your life. You have a link to the thought patterns that make up your job or occupation. And you have a link to the thought patterns of your hopes, dreams, desires and fears. Each one of these links will draw energy from your body through Yesod. The stronger the emotional ties, the more energy it takes to sustain that link. Positive thoughts and feelings will draw the energy toward you, whereas negative thoughts, emotions and feelings will draw energy away from you. It is at Yesod that you can begin to learn to control this flow of thought energy. You can begin to evaluate where your thoughts are centered and to sense how much energy the objects of your thoughts are drawing *from* you.

Each of the 10 Sephiroth corresponds to specific parts of the body. At Yesod healing for illnesses in the areas of the breast, lungs, ovaries, menstruation and stomach can be completed. Managing the thought forms relating to these parts of your body can change the flow of energy and bring about a change in your usual field of experience in these areas. This can initiate the healing process.

Yesod is the consciousness level at which true independence can be realized. Many people are afraid of any display of independence in their life because of a lack of confidence in their abilities and beliefs. If you need more confidence in your life this is where you would concentrate your efforts. How well does intuition work for you? Do you trust your own intuition? If you begin to trust your intuitive insights more, they will begin to manifest more strongly for you.

Psychic abilities also manifest at this level of consciousness. Astral Travel is a method of moving your consciousness instantly from one geographical place to another through the astral plane using your astral body. Most of us have done this at one time or another while in the Dream State. If you dream of being someplace that really seemed real after you awoke, it was probably the result of an astral trip. Some people have learned how to do this consciously by separating the astral body from the physical body and projecting it to a place they desire to be. When they wish to return to their body they will be snapped back into the body by the silver cord which connects the 2 bodies together.

Clairvoyance and Psychometry are psychic phenomena that manifest at this level of Consciousness. Clairvoyance is the ability to see things from a distance. It might be a scene from the past or the future. It could even be a scene from the present. Scenes from the future are often called prophecy. Police Departments sometimes use Clairvoyants to "see" what happened at a crime scene. Psychometry is the ability to sense the personality and character of someone just by touching an object that has been previously handled by that person, or something belonging to that person. Police Departments also use these people to gain useful information about unknown perpetrators of a crime scene. You can learn how to manage these two "psychic gifts" at Yesod.

Telepathy and Intuition are two more psychic abilities that can be developed and managed at the level of Yesod. Telepathy is the ability to pick up on someone's thoughts. The more you concentrate on the thought, the easier it is for a telepath to intercept it. Intuition is an ability that most people have to a greater or lesser extent, and involves the sudden awareness of an event that has, or will take place. Most people question or doubt their intuition and later discover that it was true, partially or in full. By learning to trust your intuitive thoughts you can strengthen the ability to use your own intuition.

The *virtue* at the level of Yesod is *independence*. This is where you will eventually begin to take *personal responsibility* for your own spiritual advancement. At that point in your life you will no longer be dependent on the family, tribe, group or religion to define your statement of beliefs. You will begin to search for more knowledge outside of the tribal parameters and to evaluate tribal or religious beliefs in light of the newly acquired knowledge. Becoming independent is the path to spiritual advancement.

The *vice* at the level of Yesod is *idleness*. An automobile or a train engine at idle may sound nice, but it is not going anywhere. So it is with an idle soul. When your soul becomes stagnant, no advancement is taking place. You feel content with your present spiritual attainment and you virtually stop learning anything of spiritual value. It is at this point that independence loses its desirability. If you feel that idleness has overpowered you, go to your internal Tree of Life and visit the level of Yesod.

The name of God at the level of Yesod is *Shaddai El Chai*. It means "Almighty Living God." A person at this level of advancement sees God as the most powerful Being in the Universe. All other Gods are inferior and subservient to the Living God. Using the mantra "Shaddai El Chai" can bring the power of the Almighty Living God to aid you in your current endeavor.

Gabriel is the *Archangel* at this level and is called "chief of guard over paradise." The *Cherubim* are the class Angels present at this level and they are the keepers of the *celestial records.* Yesod is the level where all of the information about you is stored. These Celestial Records, sometimes called the Akashic Records, are actually all of your thoughts, desires, feelings, mental pictures, and emotions stored away for future use. All of this information will provide the thought form or framework for your next reincarnation. It even provides the evidence for your level of spiritual attainment in this lifetime, which will be passed on to the next.

The Bible speaks of the celestial records and their purpose in this manner from **Revelation 20:11, "And I saw a great white throne, and him that sat on it, from whose face the earth and the heaven fled away; and there was no place found for them. And I saw the dead both small and great stand before God, and the books were opened: And another book was opened which is the book of life; and the dead were judged out of those things that were written in the books, according to their works."**

In the etheric world of Yesod, the material world and its atmosphere (heaven) are no longer visible. It is the celestial records or "the books" that give life to "the dead." "The books" provide the evidence of the level of spiritual attainment these people had attained while upon the earth. They are judged according to their works. Every thought you think is imprinted in the celestial records. Every action you take is evaluated. As we learned in a previous chapter, our purpose in this life is to learn to balance the qualities of judgment and mercy in our relationships with other people. Our success or failure in achieving our purpose is monitored by Divine Thought at the consciousness level of Yesod.

One more psychic ability that needs to be revealed is the

ability to access information from the celestial records. Some people can sense the entire character of an individual with nothing more than the person's name as a reference. From that information they can give an accurate description of that person, including they way they think and relate to other people. They can also pick up on someone else's deepest hidden secrets, known only to that individual. Be aware that nothing you do, or feel, or think, can be hidden, since it is forever available to anyone who has the ability to access the celestial records from the level of Yesod.

Yesod is the gateway to all higher levels of consciousness and is where all of the energies that flow down from those levels are organized for use on the material plane. Now is the time to organize and . . . .

awaken your energy forces.

Yesod #9

1. Yesod is where all energies are organized for use in the material plane of existence.
2. We have an etheric body, which energizes the material body and gives it the attribute of life.
3. We also have an astral body that carries and protects our consciousness.
4. Psychic abilities can be awakened and manifested at this level of consciousness.
5. The *virtue* of Yesod is the awakening of intellectual and spiritual *independence.*
6. The *vice* of Yesod is *idleness* and *complacency.*
7. Yesod is the level of the *Akashic Records* or the Celestial Records of every human being.
8. Every human thought, emotion, feeling and action is stored and evaluated by Divine Thought.

Chapter 8

HOD #8 and NETZACH #7

Hod is located on the Left Pillar, the Pillar of Judgment. This gives it a feminine potency, which manifests as power and glory. Glory is a condition of magnificence and splendor, or radiance and beauty. The essence of emotions and instincts from Netzach take form in Hod and manifest into action. Hod is the seat of the intellectual powers of man. It is the center of the faculty of reason. Reason is the ability to understand thought forms and to process those thoughts so as to draw a conclusion. It is at this level of consciousness that thought forms are empowered to begin the *transmutation* from thought energy to motivational energy.

The healing energies available at this level affect the hands, the vocal cords, the thyroid, the nervous system, the solar plexus, and the respiratory system. This is also the area of consciousness where *mental* problems manifest. So Hod is the area where you would concentrate to heal any problems associated with those areas of your body.

Hod is the sphere of *magic*. The art of magic depends on the ability to create and hold mental pictures in your mind. At this level of consciousness, thought forms are transformed into pictures or visual concepts. These mental images are necessary before anything of substance is created or manifested. Nothing can be created from a thought form alone without first becoming a mental image. Let me explain.

Suppose you decide to build a house. The thought, "I will build a house" will accomplish nothing until it is first reduced to a mental image. What kind of a house do you want? How large? What kind of rooms? What will the interior look like? What kind of walls, floors, ceilings, windows? What kind of construction materials will be used inside and outside? What kind of color coordination will be used? How will the landscaping look? Do you see a mental picture of all of these factors and more? Good!

The next step is to put all of these mental images down on paper. This is called a *blueprint*, which is used to instruct the general contractor exactly how to put this house together. If your mental images were clear and concise, and were accurately described on the blueprints, then the completed house should look exactly like the mental images you had in your mind.

Everything you see in this material world, including your own body, was manifested through this same identical process. The level of consciousness of Hod is where the mental images are formed. The level of Yesod is the Cosmic equivalent of a blueprint. The mental images are given three-dimensional form using etheric and astral substances as a medium. Because of the three-dimensional effect, the level of Yesod acts as both the blueprint and the general contractor. The material substance that you see with your own eyes then coalesces upon the astral and etheric substrate. This Cosmic Principle can be duplicated by the human mind as well. A Magician is one who has learned how to create and hold a mental image firmly in his mind long enough to form a holigramic blueprint solid enough to materialize before your very eyes. When he does this he taps into the consciousness level of Hod.

Hod is the level of consciousness that you can begin to use to imprint your will upon the substance of the astral world so as to create greater manifestation of things and events in your physical life. Techniques of mental imaging or visualization will be of greater help to you now that you know how and why it works. Even materialization and dematerialization are possible to those who have the power of concentration sufficient to manifest it. Other areas you can work on from the level of Hod are communications, travel, contracts, and making deals.

Hod is sometimes called the Plane of Splendor. Some of its activities or attributes are shared with its opposite, Netzach. For example, the interplay between Netsach and Hod creates extension, multiplication and force. Mental imaging is an example of extension and multiplication of a thought form. For example there is a thought form for the conception, growth and general form of an oak tree. But by extension and multiplication you can manifest millions of individual oak trees. And all of this is determined by force, which determines how the oak trees will behave in any given environment.

It is said that all of the forces of the Universe have come from Hod and Netzach. Scientists have credited the continued existence of the Universe to four main forces. These are the weak force and the strong force of the atom, gravitational force, and electromagnetic force. The weak force is the force that keeps electrons circling the nucleus of an atom without crashing into it. The strong force is what binds subatomic particles together to form the nucleus of an atom as a single unit. Electromagnetic force is the attraction that atoms, molecules and substances have for each other. Gravitational force is the attraction that massive bodies have toward one another. These forces seem to be similar in behavior, but have different intensities that have not yet been explained. For decades scientists have been searching for the GUT, or Grand Unification Theory that unifies these four forces into one understandable force. It is at the level of Hod and Netzach that these forces diversify and take on their individual missions.

The drive "to know God" is a result of contact with this level of consciousness. Divine Consciousness, or greater knowledge of God, can be achieved by acquiring a greater fund of knowledge. And then the Divine Consciousness will open up a greater manifestation of the knowledge acquired.

An example of this would be if you began to acquire more knowledge about how to obtain better health or healing. If this acquisition of health knowledge leads you to a greater comprehension of Divine Thought, then that comprehension of Divine Thought would begin to manifest a greater degree of health and healing than before. The same principle of expansion would apply to acquisition of enlightenment, better communications, or even greater prosperity. In support of this the Bible states in **Proverbs 22:17 and 29, "... hear the words of the wise and apply your heart unto my knowledge ... seest thou a man diligent in his business? He shall stand before kings. He shall not stand before obscure men."**

The Bible is full of instructions to pursue a lifetime of learning, and gaining knowledge and understanding, because it leads to greater trust in Divine Thought and as a consequence, greater manifestation of the desires of your heart. After all, Divine Thought is the source of the energy-information of your thought forms.

The *virtue* of Hod is *truth*. The question has often been asked, what is truth? You see, most people think that they know the truth about a lot of things in their life, and are completely unaware that what they think is true may not be true at all. Many people live in spiritual blindness. They know as truth the thought forms that were taught to them by their parents, their schoolteachers, and their general society. And those old worn out thought forms have been around for centuries. Why? Because thought forms are so very hard to change.

Have the concepts that you have been reading up to this point in this book been easy to understand, or does it feel like you are learning about a foreign culture? The answer to that question depends upon the thought forms that you have been taught since you were a child. But rest assured that everything you believe to be true does contain a certain element of truth.

Every religion, every culture, and every philosophy is based on a certain amount of truth. Some contain more truth than others do. We as human beings have the capacity to change. We are expected to search for greater truth and when we have found it, we are expected to change our beliefs and our habits to conform to our newfound truths.

The Tree of Life is composed of 10 levels of consciousness,

each of which is accessible by the human mind. Each level of consciousness adds another layer or facet of truth to that which you already possess. But it is important that you discard irrelevant and useless beliefs and thought forms that no longer conform to the greater depth of truth that you find yourself capable of accepting into your mind. The importance of truth becomes more apparent at this level of consciousness. The virtue of truth means that you must be honest about what you believe, and you must conform your life to your beliefs.

The *vice* is of Hod is *dishonesty*. Many people find it difficult to be honest in their relationships with other people. Most often this is a result of being dishonest with themselves. Many times a person cannot stand to look him/herself in the eyes through a mirror because what they see is an image that does not conform to the truths that they know. Dishonesty is a trait that can be worked on at this level. This is important because dishonesty creates an energy drain on the entire energy system that can become debilitating if left long enough. This is where the saying, "*be true to yourself*," becomes so very important.

James 1:8, **"A double minded man is unstable in all his ways."** This instability is a result of dishonesty to oneself and to others. Using the Hod level of consciousness you can begin to sense falsehoods and deceptions all around you. In this way you can develop a single-minded approach and a stronger focus of your own energies.

The name of *God* in Hod is *Elohim Tzabaoth*, which carries the meaning of "God of Hosts Ruling the Universe in Wisdom and Harmony." This is the consciousness level of Divine Thought that deals with Science and the unfolding of knowledge in world consciousness. Science means "knowledge," so this is the level at which science and spiritual knowledge should merge into one.

The *Archangel* you might encounter at this level of consciousness is *Michael*. He is called the Prince of Splendor and Wisdom. He is also known as the Great Protector, and it is said that he brings to us from God, the gift of patience. This is summed up in Romans 15:4-5, which states, **"Whatsoever things** were written aforetime were written for our learning, that we through *patience* and comfort of the scriptures, might have hope. Now the God of *patience* and consolation grant you to be likeminded one toward another . . . "

Each of the 10 Sephiroth has a *realization of conception in symbolic forms*, which is a general description of the confines of that level of consciousness. It covers not only the consciousness levels of the human mind but also the levels of the structure of the Universe, itself.

MALKUTH means *The Kingdom*, which is the totality of the entire creation. It is the work, and the mirror, of YAHWEH, Himself, and the proof of Supreme Wisdom. It is the enigma, or puzzle, that has as its answer, "YAHWEH."

YESOD means *Foundation*, the basis of all belief and truth. It is the absolute toward which all Philosophy and spiritual principles are directed.

HOD is *Eternity*. This is the triumph of mind over matter, which ultimately is the victory of life over death.

This idea is echoed in the Bible in I Corinthians 15:54, **"So when this corruptible shall have put on incorruption, and this mortal shall have put on immortality, then shall be brought to pass the saying that is written, 'Death is swallowed up in victory. O death where is thy sting? O grave where is thy victory?'"**

HOD is the Sephira on the Pillar of Judgment of the Tree of Life that is balanced by NETZACH on the Pillar of Mercy. Both Hod and Netzach work with the energy of thought forms. Netzach is the masculine energy that expresses itself as Victory. Victory can only be described as "the vision of beauty triumphant." Next time you see someone who has won a great victory observe that person's face. What do you see? You see a radiance of beauty that beams from the person's eyes and grin or smile. What you are seeing is an abundance of radiating, triumphant emotions. But it is emotion that has not been crystallized into form. It is emotion that touches the heart.

Netzach is the level of survival intuition. This is where self-esteem and self-respect are attained. These are qualities of the soul that lead to acceptance of responsibility, not only for yourself, but for others around you. These are qualities that are sorely lacking in many people today. Why? It is because low self-esteem is caused by the sense of a lack of knowledge. When one gains a sense of knowledge and understanding then one's feeling of importance rises. From this should flow a greater sense of responsibility and leadership. But this is not taught in the schools today to a majority of its students. This is why so many students in school have low self-esteem and a tendency toward extreme violence toward other students and their teachers. They have no respect for themselves and hence no respect for others. They have not been taught to be responsible for their own actions and for the welfare of others. It is no wonder that violence toward others is increasing in our world today.

Functioning at this level of consciousness can lead one to be alert to the negative energy and actions of others. Have you ever had the sense that someone you have just met did not have your's, or someone else's best interests at heart? You may have seen it in their eyes, or their face, or just felt it not knowing why. That sense of negative energy would have originated from this level of consciousness.

Another manifestation that originates here is endurance that goes way beyond the capacity of the physical body alone. We have all marveled at the reports of someone who lifted a 2,000-pound car off the body of a family member trapped underneath. This would normally be impossible without breaking bones and snapping joints. But energized from Netzach, the entire body system is temporarily transformed into conduit capable of controlling a sudden burst of power, triumph, and victory. The person is often in an altered state of consciousness, reporting that they didn't know, or couldn't remember exactly what they had done. Martial Artists who break bricks with their head or knock someone down with a fast punch that does not actually hit the body, or perform other feats of strength have conditioned their minds and bodies to use the energies in Netzach.

The healing energies of Netzach are in the realm of

complexion, hair, skin, and conditions that contribute to physical attractiveness. The kidneys and the reproduction system are also ruled by Netzach, as well as the Thymus gland.

When you combine the healing energies of Netzach's partner Hod, with those of Netzach, you will notice that the nervous system, immune system and reproductive system are all controlled at this level. This level of consciousness also ties the human mental condition to the body's physical condition and appearance. This knowledge is fueling the belief that the condition of the physical health of the body is directly related to the mental thinking of the mind. Successful Alternative Health Systems are springing up all over the world based on these concepts of the relationships of body, mind, and spirit (energy).

Netzach is the level of energy of archetypal ideas that have not yet been expresses as forms. An Archetype is the perfect idea, or ideal. But the human mind cannot fully attain the ideal in all of its perfection and universal harmony. For that reason thought forms are evolved that the human mind can grasp. And the thought forms at this level are those of the group mind.

When you think of the word "marriage," a picture will come into your mind, which probably includes scenes of a wedding. This picture, or image, comes from the world of Hod. You will also have a sense, or a feeling of what constitutes an ideal marriage. That feeling comes from the Archetypal thought form of "marriage."

There are also many *group thought forms* emanating from the world of Netzach. These are formed by various groups and cultures from the harsh realities of the rules, regulations and expectations involved in their concepts of the ideal marriage. When you think of your own marriage, you have left the group thought form for your own individualized thought form concerning marriage. The same holds true for all other thought forms. These thought forms exist throughout the Universe and your mind is able to tap into them anytime it chooses. The level of Consciousness of Netzach is densely packed with thought forms of the group mind. A group mind can be as small as 2 or 3 persons or as large as a national or even global identity.

Human instincts and emotions arise out of Netzach. However, instinct and emotion cannot be manifested by the material body without the creative power of the intellect. And conversely, intellect cannot manifest itself without the thought forms arising out of instinct and emotion. This is why there is always a constant interchange between Netzach and Hod. You see, Hod is the seat of "intellect" whereas Netzach is the seat of emotion and instinct. This interchange of intellect and emotion is part of the make-up of *Ruach*, which is that part of the soul by which you identify yourself.

Netzach is the level of consciousness from which "desire," or emotion emanates. A Theosophical viewpoint is that the desire, emotion, or instinct emanating from Netzach can be entirely dominated by *Nephesh*, the lower nature of the soul, or it can be controlled by *Neshamah*, the Divine aspect of the soul. Once instinct or emotion is energized by Nephesh or Neshamah, it can exert a strong influence over the "intellect" of Ruach.

Can you see why it is so difficult to change a habit or a way of thinking (old thought form)? Your intellect may see the need for a change, but it cannot act without the emotion. On the other hand, the old habit or thought form comes to the intellect from the seat of desire, already charged with emotion. The only way that a habit or an old thought form can be changed is to somehow charge the new one with even greater emotion. One way this can be done is by gaining control of emotion and instinct from *Tiphareth*, the energy level of consciousness beyond Netzach and Hod.

Let's say you decide, "I am going to quit smoking, or I am going to lose weight beginning today." At the end of the day you realize that you did not accomplish your desire. Why? It was because the desire to continue the old habit was stronger than the desire to establish the new habit. The old habit had emotions, feelings, attachments, memories and comfort to give it strength. Each one of those is an energy attachment that would need to be cut off in order to reduce the power of the old desire. To be successful the new desire would need to be energized with excitement, enthusiasm, devotion, and other forms of stimulation to give it enough energy to over ride the old. Strength of desire determines which intellectual decisions you make in your life will be successful.

Personal development in the area of Netzach and Hod can center on challenging the group mind. Since this is the level of group thought forms, the thought forms of all of the groups you participate in are gathered here. The ideal thought form is here as well. Catching a glimpse of the *ideal* and comparing it with the actual often presents one with a decision. Will you stay with the old flawed group and challenge it to change, or will you leave the old group in search of one that is closer to your concept of the ideal? This is where personal decision making can be strengthened and improved.

When you are in a group, most of your decisions must take the rest of the group into consideration. As you develop greater decision making abilities, you also increase your own personal responsibility. If you gain a leadership role in the community or group, then the people in the group will follow you. However, if the people cease following you because they "fear" the direction they see you leading them, then you will either need to turn back, or leave the group behind and trudge on alone. From this point onward the group will no longer be your conscience or your guide. You will now have begun a lifelong search for a stronger connection between yourself and Divine Thought.

The *virtue* of Netzach is *unselfishness*. Unselfishness is an emotion of a higher nature. When the higher aspects of the soul are in control of the emotions these emotions manifest themselves as unselfishness and love. This is where the purpose of the soul begins to change from "receiving for the self alone," to "receiving so as to share with others." However, the sharing will take place at a higher spiritual level than before because you will be operating more from the level of *Neshamah*. Netzach is also the source of sexual love. The energies that flow from sexual love can be of a very high nature.

The *vice* of Netzach is *lust* and *impurity*. If the energies at this level become excessive and unbalanced they will lead to lust, and not just sexually. When thoughts become impure, lust for almost anything can increase. The desire to receive (or take) for one's own selfish gratification becomes flagrant and the soul falls under the influence of Nephesh. Netzach is the level of consciousness at which you can work on these vices, to rise to the unselfish desires of Neshamah.

From Netzach you can obtain understanding and power in personal relationships. Many times relationships are strained to the breaking point simply because of misunderstandings. By working at the level where emotions emanate, you can gain immediate understanding and resolve the root cause of the misunderstandings. You can strengthen your relationships with others simply by putting more positive energy into them.

Idealism is also a result of contact with this level of consciousness, because Netzach is the source of all *ideals*. A true visionary is one who is in contact with this level of the Tree of Life.

The name of God at this level is *Yahweh Tzabaoth*, which is "*God of Hosts*." This means that the God energy at this level controls everything in its diversities and varieties. The energy of God at this level can assist you in expressing your wide-ranging emotions in the most positive and beneficial manner.

The Archangel of the Sephira of Netzach is *Haniel*. She is known as the archangel of *love* and *beauty*. She also rules over the arts, which are expressions of love, beauty, emotion and idealism.

The realization of Conception in Symbolic Form is *Victory*. It is the triumph of intelligence and justice. Justice is the proper exercise of power and authority to maintain what is right, equitable fair and just.

Pathworking is a method of using the Tree of Life to obtain its benefits in a balanced form. A vice is usually the result of excessive energy, or energy out of balance. Pathworking achieves balance by working with triads or 3 Sephiroth at a time. Using a Sephiroth from each of the 3 pillars helps to achieve balance. When you visualize the superimposed Tree of Life you can access the energies of any one Sephira through a triad, such as Malkuth-Hod-Netzach, or Yesod-Hod-Netzach, or Tiphareth-Hod-Netzach.

By visualizing the triad and concentrating on the specific quality you wish to strengthen, you draw the energies of that specific quality in a balanced form.

The interaction between Hod and Netzach is what fuels your intellect and powers your thoughts and actions. All of your emotions and desires emanate from this interaction and determine the quality of your thought patterns. In order to change a bad habit or thought pattern, you must replace it with a new habit or thought pattern and . . .

Become empowered with emotion and desire.

Hod #8, and Netzach #7

1. Hod is the seat of intellectual energies.
2. Hod is where thought forms are transformed into pictures, or images in the mind.
3. The drive "to know God" begins here.
4. Knowledge of God increases the effect and comprehension of any other knowledge acquired.
5. The *virtue* of Hod is *truth*. The *vice* of Hod is *dishonesty*.
6. Netzach is the seat of emotions and and instinct.
7. Netzach is where thought forms originate based on Archetypes.
8. Instinct and emotion cannot be manifested without the intellect.
9. Intellect cannot manifest itself without the thought forms emanated from emotion and instinct.
10. The *virtue* of Netzach is *unselfishness;* the *vice* of Netzach is *impurity* and *lust*.
11. You can obtain balanced energy from the Tree of Life by using triads.

Chapter 9

TIPHARETH #6

Tiphareth is the Sephira of beauty and harmony. In the Zohar it is written that Tiphareth is **"the highest manifestation of ethical life, the sum of all goodness; in short, the Ideal."** In Greek the meaning of ethics is basically living the ideal. Tiphareth is located on the Central Pillar of Equilibrium, and its function is to provide equilibrium, balance and harmony to everything in the Universe. Tiphareth also brings together all of the elements needed to bring your body, mind, and spirit into perfect balance and harmony. The soul achieves balance at this level as *Neshamah*, *Ruach* and *Nephesh* harmonize to move you on to achieve your destiny.

All general health problems can be affected at Tiphareth. Disorders of the back and spine can be improved at the level of Tiphareth. Things pertaining to the heart, blood, and circulatory system are also at this Sephira. The heart affects the entire body because it is responsible for pumping the blood, or fluid of life to every organ and every cell of the body. The blood cleans away all debris and trash and replaces it with new sustenance and nutrition. When body organs break down and the circulatory system becomes clogged, this energizing process becomes slower and weaker. When the heart and circulatory system become overworked in the simple process of trying to sustain the bodily functions, the entire energy system loses its stamina and the entire body health declines. This can all be turned around at Tiphareth, because this is where healing energies manifest. This is also where the ability to heal others originates.

Tiphareth is an emanation of Chesed and Geburah. These are the energy levels of Justice and Mercy. Tiphareth is the beauty, the balance, the equilibrium, the harmony, the mildness, or the clemency that results from the blend of Justice and Mercy. Tiphareth is also a point of transition because it is connected to all of the other Sephiroth. Because of its unique position it can *mediate* between any two of the remaining energy levels or Sephiroth.

Tiphareth can also mediate between *all* of the other levels of consciousness and create a perfect blend of them all. This is why it is called Microprosopan, or *small personification*. It is also called *small face* because it is a reflection of the higher truth of the Endless Light of Kether. The perfect energy of Kether is too potent for our finite minds, but Tiphareth can filter it through Justice and Mercy so that our mental processes can grasp it.

Tiphareth the Mediator

What is beauty? Can you describe or explain beauty? Beauty is an energy level beyond the 5 senses. Beauty is something that is felt from the center of your being. You may feel beauty in something that you see, or hear, or touch, or smell, and that feeling comes from the center of your being. For example, you might look at a scene, or listen to a piece of music 100 times, and then suddenly one time you feel a tremendous *sense of beauty* that you cannot even describe. When you are centered in Tiphareth you can "*see*" beauty in anything or everything.

There is a saying that "beauty is only skin deep." I beg to differ. What that statement means, in my opinion, is that symmetrical proportions, curvaceous lines, or being pretty is just skin deep. Beauty is something different. The phrases "*inner beauty,*" and "*beauty radiates from within,*" gets closer to the Kabbalistic truth of what beauty really is. It is an energy that radiates from within everything and everyone, that can be sensed under the proper conditions.

Beauty is an essence that refers itself to the ideal or absolute beauty. Harmony is an essence that designates absolute harmony. Equilibrium is an essence that emanates from absolute equilibrium. These are not 3 distinct essences but just one. In the Endless World, and in Kether, these are one and the same essence. In Tiphareth these are 3 essences *blended* into one. Our minds can comprehend nothing except by division so we perceive 3 different phases of this one essence. You see, beauty is a harmonious blend of various energies as detected by our own senses. But unfiltered by our senses this energy is a Unity that surrounds us and permeates our being continually. The Bible expresses this diversity in Unity as well. The Psalmist said in Psalms 96:5-7, **" . . . (YAHWEH) made the heavens. Honour and majesty are before him; strength and beauty are in his sanctuary (dwelling place). Give unto YAHWEH O ye nations of the people, give unto YAHWEH glory and strength."**

A Kabbalist would understand this scripture to say that Divine Thought created the levels of consciousness we call "*the heavens.*" Honour (Hod in Hebrew) majesty and glory all relate to Hod, a Sephira we have already studied. Strength, which is Power, relates to Netzach. These are all energy levels from Divine Thought that we are to emulate, strengthen and use as we go about our daily lives. However, in YAHWEH these are all One.

The Zohar speaks of *The Mystery of the Balance*, which is the secret of the Universal Equilibrium. When you take Infinite Divine Wisdom and balance it with Infinite Divine Power, you obtain Infinite Stability of the material Universe as well as the unchangeableness of Divine Law. This dual principle manifests as truth, precision, justice, impartiality and rightness. Moreover, all of the laws of *nature* (mathematics, chemistry, physics, electro-magnetics, electricity, gravity, ecology, and biology) stem from Absolute Divine Law. Equilibrium between Infinite Divine Justice and Infinite Divine Mercy produces Infinite Divine Equity, moral harmony, and beauty. This equilibrium allows for the flawless perfection of Divine Thought to endure the faulty imperfections of the material world.

Tiphareth can bring harmony and unity to all things. When the energies of Tiphareth are in balance then everything below will be in balance as well. The emotions of Netzach will be in balance and harmony, energizing positive productive performance. The quality of the Intellect and the thought patterns of Hod will be centered on the good of others rather than on attaining for one's self. All of this is because balance, beauty, harmony, unity, and equilibrium are centered in Tiphareth.

The *virtue* of Tiphareth is *Devotion to the Great Work*. When one reaches the level of Tiphareth one's focus begins to change from self-serving to a desire to serve others. This drive to serve others usually results in finding an avenue of service or "A Work" with which one can be of benefit to others. Those who are balanced in their "Devotion to the Great Work" will usually be successful in their endeavors, because of their single-minded dedication. For many, glory and fame are even possible because "the Great Work" can cover a very broad range of activities and services.

The Great Work for many dedicated people centers on issues of "life." These dedicated workers want to help people solve problems in their personal lives so that they might physically, mentally, emotionally, intellectually or spiritually advance. Tiphareth is the center for matters that concern "healing." So the healing arts in all of its forms are usually practiced by individuals at this level.

Awakening of the "Christ Consciousness Within" also occurs at this level. As you will soon learn, everything in Tiphareth is centered on "Christ Consciousness" and the "Messiah Concept" or "The Savior."

The *vice* at this level is *false pride*, which happens when this level of Consciousness becomes unbalanced. Let me tell you that it is very easy to become unbalanced in Tiphareth because of the pull of all of the other Sephiroth on Tiphareth. False pride is evident in many people who have achieved the pinnacle of success in their field of "the Great Work," and some have tumbled because of it. No matter who you are or what your line of work, always be aware of false pride, because it will indicate to you that you are out of balance with the Absolute Power of the Universe.

Devotion to the Great Work is for the purpose of Soul advancement. One of the joys of understanding the Great Work is watching the Universe unfold as it should. Nothing that happens in this Universe, and in any individual's life, is contrary to the Will of Divine Thought. However you must be mindful that Divine Thought allows for great latitude of free will so that you will have ample time to discover and align yourself with that Divine Will. Divine Thought is immutable and unchanging. Whenever you do not like what is happening in your life, that is an indication that it is time to assess where you stand in relationship to the Will of the Divine Thought of the Universe.

You are one entity although you have a dual nature. You have one nature that desires things for yourself, and a second nature that desires to share things with others. You begin to attain your purpose in life when you are able to harmonize and balance these two natures. When your life is in harmony and balance only then will you begin to be truly successful.

The stars in the Heavens are in Harmony with the laws, forces and spiritual qualities of the Universe. All of the forces and balances of Nature harmonize themselves with the great Harmonies of the Universe. But Man, and only Man, has the ability to choose whether or not to harmonize himself with the equilibrium of the Universe. *Ruach,* the Ego which is *you,* must choose whether to Harmonize with the spiritual qualities of *Neshamah,* or be ruled by the self-centered *Nephesh.*

The Kabbalah gives many different titles to Tiphareth. Some of these are; The Lesser Countenance, The King, The Man, Adam, The Son, and The Sun. Tiphareth also has as symbols, A Child, A Sacrificed God, and the Mystery of Sacrifice. Each one of these will be explained from the Kabbalah and from the Bible.

The Lesser Countenance is a title given to Tiphareth because it is a reflection of the qualities of Kether, the first Sephira, which is called "The Greater Countenance." All of the energies and powers of the various levels of the Universe are emanated from Kether. These energies are so powerful that they cannot be comprehended by the human mind. It is a unity so powerful that it cannot be divided or fathomed. As the energies emanated down the Tree of Life each Sephira retained its own qualities and energies. When Tiphareth blends, balances, harmonizes and reunites all of these energies into one equilibrium, it is a *reflection* of the energy that remains in Kether. But it is a harmonized energy that can be separated and accessed by the human mind and intellect.

The King is a title of Tiphareth because it refers back to Emanations that occurred before the present Universe came into being. 7 kings were emanated and later destroyed because there was no equilibrium. This occurred before Adam Kadmon became both male and female. The 7th king, Regnum, became the substance out of which the present Universe was emanated. For that reason Regnum was called *"The Cornerstone."* When Adam Kadmon became male and female balance was established and the emanation of the Universe proceeded according to the Cosmic Plan.

The Man is a very important title of Tiphareth because it signifies the whole purpose of the Universe. In the very beginning Man had the Word, and that Word was from God. Man was created pure; and God gave him *truth*, as He gave him *Light*. He lost the *truth* and discovered *error*. He wandered deep into darkness. And in thick darkness he remains to this very day.

Here is how it happened. In the Beginning, *Divine Thought* was but *One SOUL*. It was the all enduring Infinite. It had this thought: '*I Create Worlds:*' and instantly the Universe, and the laws of harmony and motion that rule it, came into being, including birds and beasts and every living thing but Man. A second thought was expressed: '*I create Man, whose Soul shall be in my image, and he shall rule.*' And out of the soil Man was formed with senses, instinct, and a reasoning mind! And yet not Man! Just an animal that breathed, and saw, and thought; until a spark of life from The One Soul penetrated the body and Man became a living soul. Man the Immortal! Before the world grew old, the primitive Truth faded out from the soul of Man. Then Man asked himself, '*What am I? And how did I get here and where am I from? And where will I go?*' And the Soul, searching itself, endeavored to learn whether that "*I*" was just mere matter.

Tiphareth brings balance to the entire Universe. It also balances the Soul within Man. The Man represents Tiphareth because Man was created to bridge the gap between Heaven and Earth. Man once had the truth about who he is, but lost that knowledge. Even today the same questions are asked, 'Who am I, where did I come from, and where am I going?' So Man is now responsible to reclaim that knowledge and put it to use. This is explained in Psalms 8:3-6, **"When I consider your heavens, the work of your fingers, the moon and the stars, which you have ordained, what is Man that you are mindful of him? And the son of man that you visited him? For you have made him a little lower than the angels, and have crowned him with glory and honour. You made him to have dominion over the work of your hands. You have put all things under his feet."**

Each of the Sephiroth, or levels of consciousness, of the Tree of Life are ruled by a class of Angels and Archangels. Man was designated to have dominion over the material world. But Man also has access to each of the other consciousness levels, unlike the angels. This is why Man is the bridge between Heaven and earth. It should also be noted that the phrase "lower than the angels" is misleading because the word translated as 'Angels' is ELOHIM. As you know, ELOHIM means Deity or God. So the text should read, "you have made (man) a little lower than God." This understanding makes Angels servant to Man when he learns how to command them from any level of consciousness. Man is, therefore, responsible to bring balance and equilibrium to the Earth and all of its inhabitants.

Adam is another important title of Tiphareth because it is almost interchangeable with Man. Adam represents the first Creation that was modeled after the Adam Kadmon. As you learned in an earlier chapter, Adam Kadmon is the body of God, or Divine Thought. It is the image, or thought form, from which the Adam of clay was created and endowed with Soul. It is the responsibility of each Soul within the Adam of clay to return to the Unity of Adam Kadmon. When Adam Kadmon was fashioned into male and female, equilibrium was introduced and everything flowed out from the two, which are really one. No relationship other than male and female can establish equilibrium. This was all symbolized by the Genesis account of the creation of Adam and Eve. They were created in the image of Adam Kadmon, first as a generic individual, then separated into male and female. The male and female, when married then learn to operate as one. This is revealed by the statement in Genesis 2:24, **"Therefore shall a man leave his father and his mother, and shall cleave unto his *wife*, and they shall be *one flesh*."**

The Son is a title of Tiphareth. This goes back to the very beginning of time, to the emanation of Adam Kadmon. The very first Emanation from the First Cause was Divine Thought Thinking. Divine Thought began by conceiving the Universe. This thinking and conceiving was the Spirit, or energy of Emanation. This was

The Word, or *Logos*, which was the Creative Utterance of Divine Thought. Logos means the spoken word including the thought motivating the word. The Logos, or Creative Utterance of Divine Thought is also known as the First Begotten Son. Many sons (a term that includes both sons and daughters) have emanated from Divine Thought. But the Logos was the first and preceded the others, which we referred to earlier as Sparks of Life. John 1:1-4 alludes to this by saying, **"In the beginning was the Logos (Word), and the Logos was with God and the Logos was God. The same was in the beginning with God. All things were made by him, and without him was not anything made that was made. In him was life, and the life was the light of men."**

The light of men, of course, is the indwelling Spark of Life. The Word is the *First Begotten;* not the first created Son of God.

The Adam Kadmon is called the *Only-Begotten Son.* This is because Adam Kadmon is the pattern for the Universe (macrocosm) and for Man (microcosm).

John 3:16 begins, **"For God so loved the world that he gave his only begotten Son . . . "** This statement does not refer to the "virgin birth" of Yahshua (Jesus), as most people believe, but to the Logos within him. This is made more clear in John 1:14, **"And the Word was made flesh and dwelt among us, (and we beheld his glory, the glory as of the only begotten of the Father,) full of grace and truth."**

Peter, James and John had the opportunity to see the transfiguration of Yahshua into his body of light. John believed that the glory (brightness) of the only begotten, was the Logos, or The Word in its body of Light. This was the very same Logos, or Word, that created the universe.

This is acted out in Genesis Chapter 1 describing the story of Creation (Emanation). In the beginning, the earth was formless and void. First Cause poured out His Spirit which flowed across the waters and every Word He verbalized came into being. The very first Word spoken became the Only-Begotten Son, which then became the agent of creativity. **"In the beginning God created the Heaven and the earth. And the earth was without form and** void. And darkness was upon the face of the deep. And the spirit of God moved upon the face of the waters. And God said, 'let there be light' and there was light."

Tiphareth takes the energies that come from above and puts them in perfect balance and equilibrium. This ensures that the final emanations below will be beautifully balanced. And this is the function of the Son of God, or the Word, which resulted from the Male and Female Principles of emanation. That is why The Son is a title of Tiphareth.

The Sun is a title of Tiphareth because it represents the energies that give life to the earth. All of the ancient religions regarded the sun as the soul of the Universe, and they believed that the sun traversed throughout the Universe in its daily routine. This is symbolized in the ancient book of Job 22:12-14, **"Is not God in the height of heaven? And behold the height of the stars, how high they are. And you say, how does God know? Can he judge through the dark cloud? Thick clouds are a covering to him, that he sees not. And he walks in the circuit of heaven."**

The sun represents the One and Only God. The word in Latin for the sun is *SOL*, which means solo or One. This symbolizes Unity and Harmony. Psalms 84:11 declares, **"For the Lord God is a sun and shield; the Lord will give grace and glory; no good thing will he withhold from them that walk uprightly."**

One of the important aspects of Tiphareth is that it is the source of everything relating to the *heart*, to *health* and to *healing*. Proverbs 3:5-8 and 4:20-23 in this regard says, **"Trust in the Lord with all your heart and lean not unto your own understanding. In all your ways acknowledge him and he shall direct your paths. Be not wise in your own eyes, [respect] the Lord and depart from evil. It shall be health to your navel and marrow to your bones . . . My son, attend to my words and incline your ear to my sayings. Let them not depart from your eyes; keep them in the midst of your heart. For they are life to those that find them and health to all their flesh. Keep your heart with all diligence, for out of it are the issues of life."**

The wisdom of King Solomon shown in the Proverbs above, demonstrates Kabbalistic Principles. First of all it indicates that when we align our thinking and belief system to that of Divine Thought, it will have beneficial effects on the state of our health. The word *navel* means umbilical cord, which is the source of nourishment from the mother to the baby. The analogy is that in understanding the direction of God we allow the life-giving energy to flow from God into our physical bodies. As we know from medical science today, *marrow* is the source of the life-giving properties of the blood. So by opening up your understanding of Divine Thought, the energies from Tiphareth can flow directly to the very source of the vitality and health of your body.

We understand that the heart is responsible for sustaining (issuing) the physical life. And we also know that the heart represents our deepest and most entrenched thoughts and beliefs. This Proverb is teaching us that by gaining the knowledge, understanding and wisdom of Divine Thought, we actually affect our physical health in a positive way. Remember that *thought is energy*. And positive thought produces a flow of positive energy that sustains our physical existence. The ultimate goal is eternal existence, or Immortality.

Malachi 4:1-3, **"Behold, the day will come that shall burn as an oven. And all the proud, yea, and all that do wickedly shall be stubble. And the day that comes shall burn them up, says the Lord of hosts, that it shall leave them neither root nor branch. But unto you that [respect] my name shall the Sun of righteousness arise with healing in his wings, and you shall go forth and grow up as calves in a stall. And you shall tread down the wicked, for they shall be ashes under the soles of your feet in the day that I shall do this says the Lord of hosts."**

The healing provided by the Sun of righteousness is actually eternal life, which provides the ability to survive the heat that will destroy the bodies of mere mortals. The Immortal body manifests on the material plane but is composed of elements from a higher plane of existence.

The symbol of a Child represents Tiphareth in that it depicts a child born for the purpose of providing a sacrifice of his life for the benefit of humanity. This is also represented in other ancient religions, as were the preceding titles of Tiphareth. Since the ancient religions all had versions of these symbols it indicates that they must have been derived from a common source. And since the Tree of Life was originally introduced to Adam and Eve, it would seem that the Tree of Life would be the source of these similarities. The symbol of the child would represent the Son, or the Only-Begotten, which is the Logos. Isaiah carries this idea twice. In Isaiah 7:14-16 we read, **"Therefore, the Lord himself, shall give you a sign; behold, a virgin shall conceive and bear a son, and shall call his name Immanuel (with us is God). Butter and honey shall he eat, that he may know to refuse the evil and choose the good. For before *the child* shall know to refuse the evil and to choose the good, the land that you abhor shall be forsaken of both her kings."**

Isaiah continues this idea in Isaiah 9:6-7, **"For unto us a *child* is born, unto us a son is given; and the government shall be upon his shoulder; and his name shall be called Wonderful, Counsellor, The Mighty God, The Everlasting Father, The Prince of Peace. Of the increase of his Government and peace there shall be no end, Upon the throne of David and upon his kingdom, to order it, and to establish it with judgment and justice from henceforth and forevermore. The zeal of the Lord of hosts will perform this."**

This prophecy would not come to pass until after the land was forsaken of both of her kings and until both Israel and Judah ceased to exist. The fulfillment of this prophecy is indicated in Luke 1:30-33, **"And the angel said unto her, fear not Mary, for you have found favor with God. And behold, you shall conceive in your womb, and bring forth a son and shall call his name Jesus [YAHOSHUA—Salvation of God]. He shall be great and shall be called the Son of the Highest; and the Lord God shall give unto him the throne of his father David. And he shall** reign over the house of Jacob forever, and of his kingdom there shall be no end."

The Symbol of a Sacrificed God is the centerpiece of Tiphareth because it represents The Great Work! What is The Great Work? It is the process of restoring Man back to his original estate and ascending back into Union with Divine Thought. Ephesians 1:3-4 spells it out by stating, **"Blessed be the God and Father of our Lord Jesus Christ, who has blessed us with all spiritual blessings in heavenly places in Christ; According as he has chosen us in him *before the foundation of the world* that we should be holy and without blame before him in love."**

You learned in an earlier chapter that our "sparks of life," which fuel our souls, became soiled in the Endless World before the Universe was emanated. And we know that the Universe was emanated so as to provide the opportunity for us to restore ourselves to a state of blamelessness and holiness before Divine Thought. And this would be accomplished through the work of a Sacrificed God.

All ancient religions had a concept of a sacrificed god, who was slain for the salvation of his people. In many religions this was memorialized by human sacrifices to their gods. But since the beginning of time people knew that a God would be sacrificed to restore the balance that was lost before the beginning of time. This concept is revealed in **Revelation 13:8**, which speaks of those **" . . . whose names are not written in the book of life of the Lamb slain from the foundation of the world."**

In the days of Abraham, child sacrifice was rampant all over the world. Fear and tradition made human sacrifice binding on people and nations perpetually. In discovering the knowledge of Kabbalah, Abraham came to understand Divine Thought, and the idea of a sacrificed god. However, like many of us, his understanding was merely intellectual. To really *believe* would require emotion, and this is why Abraham received the supreme test from his "friend" YAHWEH. Abraham was instructed to go into the wilderness and offer his son Isaac as a burnt offering. When Abraham and Isaac arrived at the location where the sacrifice would take place, Isaac asked (Genesis 22:7-8), **"Father . . . I see the fire and the wood, but where is the lamb for the burnt offering?"**

Abraham answered, **"My son, YAHWEH will provide himself a lamb for a burnt offering."** Abraham then built an altar out of stone, laid the wood upon it, then tied Isaac and laid him on the wood, and actually raised up the knife to kill Isaac, when the Angel of YAHWEH commanded Abraham to stop. Then Abraham and Isaac heard these words (verse 12), **"lay not thy hand upon the lad, neither do anything unto him, for now I know that you trust YAHWEH seeing you have not withheld your son, your *only* son from me."** Then they saw a ram caught in a bush by its horns and they used it for the sacrifice.

There are people today who feel that a God who would put Abraham through a test like that is not worthy of "worship." These people also seem unable to understand how a God could authorize the sacrifice of His own Son on behalf of humanity. However, these people do not realize that without the sacrifice it would be impossible to restore the material creation back into balance. Nor would humanity ever be restored back into proper equilibrium with Divine Thought. Mankind would not get better, but worse, until the whole earth and the Universe were finally, irretrievably destroyed. It was necessary for Abraham to believe *strongly* enough so that he could emanate that belief down through the ages until a time when enough minds were capable of comprehending the sacrifice and how it would help to bring the earth and the Universe back into equilibrium. Hebrews 11:17-19 explains, **"By faith Abraham, when he was tried, offered up Isaac; and he that had received the promises offered up his *only-begotten son,* of whom it was said, that in Isaac shall thy seed be called. Accounting that God was able to raise him up, even from the dead; from where he also received him in a figure."**

Abraham came to understand the sacrifice that Divine Thought made when he sent the Word into the Universe at the time of the original emanation. As you learned in an earlier chapter, the Universe was emanated so as to correct the Bread of

Shame, which was an imbalance in the Endless World. The material Universe has given us a place to work out this problem over a period of time, but the sacrifice of a God would be necessary, and was built-in to the emanation of the Universe from the very beginning.

The Apostle Paul speaks of the restoration of the Universe in this manner; Romans 8:19-22, **"for the earnest expectation of the creature waits for the manifestation of the sons of God. For the creature was made subject to vanity, not willingly, but by reason of him who has subjected the same in hope, because the creature itself also shall be delivered from the bondage of corruption into the glorious liberty of the children of God. For we know that the whole creation groans and travails in pain together until now."**

You see, the purpose of the sacrifice of a God was to initiate the development and manifestation of the sons (and daughters) of God. The creature, or creation, was made subject to vanity or temporary existence. It is not only humanity that is obsessed with the quest for immortality, but the living Universe is also aware of this quest. Once the process of the manifestation of the children of God reaches critical mass the entire Universe will be on the threshold of equilibrium and the cosmic Tree of Life will become recentered at Tiphareth.

The sacrificed God was anticipated in the Bible and was described in this manner by Isaiah 52:13-53:12 **"Behold my servant shall deal prudently, he shall be exalted and extolled, and be very high. As many were astonished at you, his visage was so marred more than any man. And his form more than the sons of men . . . he has no form nor comeliness, and when we shall see him, there is no beauty that we should desire him. He is despised and rejected of men, a man of sorrows, and acquainted with grief; and we hid as it were our faces from him, he was despised and we esteemed him not. Surely he has borne our grief and carried our sorrows, yet we did esteem him stricken, smitten of God, and afflicted. He was wounded for our transgressions, he was bruised for our iniquities, the** chastisement of our grief was upon him, and with his stripes are we healed. All we like sheep have gone astray, we have turned everyone to his own way, and the Lord has laid on him the iniquity of us all. He was oppressed and he was afflicted, yet he opened not his mouth; he is brought as a lamb to the slaughter, and as a sheep before her shearers is dumb, so he opened not his mouth. He was taken from prison and from judgment; and who shall declare his generation? For he was cut off out of the land of the living. For the transgression of my people was he stricken . . . By his knowledge shall my righteous servant justify many, for he shall bear their iniquities. Therefore will I divide him a portion with the great, and he shall divide the spoil with the strong, because he has *poured out his soul* unto death. And he was numbered with the transgressors, and he bare the sins of many, and made intercession for the transgressors."

This prophecy anticipates that a *man* would be required to bear the essence of God and to pour out his soul for the benefit of humanity. The Sacrifice of a God is the method by which the material creation will be brought back into balance and centered in Tiphareth. This process also occurs on the individual level, which was referred to as "bringing into manifestation the sons of God." Now we can look a little deeper into the *mystery*.

The Mystery of Sacrifice is the key to truly understanding Tiphareth. The energy-information from Tiphareth fuels the thoughts and emotions that emanate from Hod and Netzach. This energy-information, when flowing properly from Tiphareth, gives beauty, balance and equilibrium to all of your thoughts, feelings and emotions. But for most people the energy-information flow from Tiphareth is not in proper balance. The energy-information that flows through Tiphareth is the source of life. Tiphareth is the Sephira of healing and health because its energies relate to the quality of life.

The energy-information of "life" is carried and sustained in the body by the blood. The ancients knew this. Human sacrifices were a way that one person could atone for the lives of many. However, that atonement did not fix the problem but just pushed the problem another month or year into the future. As the blood drained from the victim, the life, or animating soul, was offered as the sacrifice for the many. When animal sacrifices were established for the people of Israel they were told, **Deuteronomy 12:23, "Only be sure that you eat not the blood, for the blood is the life; and you may not eat the life with the flesh."**

The word *"life"* here is Nephesh, which is the soul. When an animal is killed for food, the blood is drained out of the body first, which symbolizes pouring out the animal, or animating soul back into the earth. This is reemphasized in **Leviticus 17:11, "for the life [Nephesh] of the flesh is in the blood, and I have given it to you upon the altar to make an atonement for your souls. For it is the blood that makes an atonement for the soul . . . ."**

We must always keep in mind that our purpose in life is to ascend back into unity with Divine Thought. Tiphareth is where diversity is blended back into unity. Humanity has descended to a condition from which it cannot attain unity on its own. Help is necessary to remove some of the barriers, blockages and restrictions that create the diversities that impede our spiritual progress. These barriers exist in the form of negative energy, which affects the spiritual pattern, as well as the earthly manifestation. Therefore, the atonement provided by human or animal sacrifices cannot be permanent, because it cannot purify the heavenly pattern. Only the energy-information of a God can do that, which explains the necessity for a *God Sacrifice*.

Hebrews 9:22-26, "And almost all things by law are purged with blood, and without shedding of blood is no remission. It was necessary that the pattern of things in the heavens should be purified with these. For Messiah is not entered into the holy places made with hands, which are the figures of the true, but into heaven, itself, now to appear in the presence of God for us. Nor that he should offer himself often as the high priest entered into the holy place every year with the blood of others; for then must he often have suffered since the beginning of time. But now, once in the end of the age, has he appeared to put away sin by the sacrifice of himself."

How is it possible for a God to sacrifice himself? From the point of view of Abraham's Kabbalah, how could Divine Thought provide this sacrifice?

It all began at the emanation of the material Universe. Divine Thought poured out its spirit, or energy-information, and spoke, "Let there be Light." Immediately, there was Light. This Light was the energy out of which the Universe would be manifested. This same Light was the *Only-Begotten Son* of Divine Thought. It was the result of the vibration of the spoken WORD that launched the emanation of the material universe. The WORD essentially became the Universe. The sacrifice on the spiritual level essentially began at this point, because the Universe is subject to decay and mortality, and the WORD, or Light, was now in the transitory Universe.

However, the physical sacrifice would require human form, and this is explained in **John 1:1-14, "In the beginning was the Word, and the Word was with God, and the Word *was* God. The same was in the beginning with God. All things were made by him, and without him was not anything made that was made. In him was *life*, and the life was the *light* of men. And the light shined in darkness and the darkness comprehended it not . . . That was the true Light, which lights everyone that comes into the world. He was in the world, and the world was made by him, and the world knew him not. He came unto his own and his own received him not. But as many as received him to them gave he power to become the Sons of God, to them which believe on his name; which were begotten not of blood, nor of the will of the flesh, but of God. And the Word was made flesh and dwelt among us (and we beheld his glory, the glory as of the only begotten of the Father), full of grace and truth."**

When YAHSHUA was born he symbolized "the child" born in the flesh to a woman to become a sacrificial God. He also symbolized the Man as he was called "the man of sorrows." When he received the Logos, or The WORD at the time of his baptism, he symbolized the Son of God and the Only-Begotten Son. He also symbolized the King, representing Regnum the 7th king when he was called the chief Cornerstone who would restore equilibrium to all of creation. When YAHSHUA, the Son of Man, sacrificed his life and died, he was pierced with a sword and his blood fell to the ground while the essence of the WORD was released into the Universe. I Peter 3:19 indicates that while the body of Yahshua lay in the grave, the Logos went to a spiritual plane where he was now qualified to bring to judgment spirits that had erred in the past. At the resurrection of YAHSHUA, after 3 full days of death, the physical body was dissolved and became the "Glorified body," or body of light. Then Yahshua ascended to present himself in the presence of Divine Thought as the restored Adam Kadmon, by dissolving the atoms of the "Body of Light," and the sacrifice was finally complete. This is called "The Great Work."

I Corinthians 15:45,49, **"And so it is written, the first man Adam was made a living soul, the last Adam a quickening spirit. And as we have borne the image of the earthy, we shall also bear the image of the heavenly."**

Ever since the first appearance of Adam and Eve, mankind has been unable to raise itself back up to the level of spiritual light. Man became a living "Nephesh" in which he was ruled by his human nature. The last Adam, Adam Kadmon, or the WORD, demonstrated with his physical life how to use Neshamah to balance the drive of human nature. His victory over death gave him the power to impart Immortality to others. That is what is meant by a "quickening spirit." And we now have the opportunity to attain to Eternal Life in one lifetime!

The Apostle Paul explains how we can carry on the Great Work in our lives today, **Romans 12:1-2, "I beseech you, therefore, by the mercies of God that you present your bodies a *living sacrifice*, holy, acceptable unto God, which is your spiritual service. And be not conformed to this world, but become transformed by the renewing of your mind, that you may prove what is that good, and acceptable, and perfect will of God."**

The Great Work for any individual is to become a *"living*

sacrifice." What is a living sacrifice? It is basically allowing your Neshamah, to manage your daily life. When you dedicate your life toward spiritually uplifting a struggling humanity, you are participating in The Great Work. At first you will begin by learning to manage your own spiritual life by moving beyond the tribal level of allowing others to do your spiritual thinking for you. As you become more spiritually independent of others and draw closer to Divine Thought you will expand your understanding of life around you beyond anything you thought possible. As your understanding increases and you become more balanced in your life you will feel the need and the drive to share some of what you know with others who are struggling in a sea of confusion, bewilderment and perplexity. As you begin to help others to learn more about themselves and their place in this universe, you will have begun to participate in The Great Work!

The name of God at the level of Tiphareth is *YAHWEH Aloah va Daath*, which means *God of Knowledge and Wisdom*. The Bible is full of references to the wisdom and knowledge of God, and that God is the source of all knowledge and wisdom. As you learn the Kabbalistic understanding of the relationship of the Universe to mankind, you begin to sense the breadth and scope of that perfect knowledge. Drawing on that knowledge can be accomplished by centering on Tiphareth in the Tree of Life.

The Archangel is *Raphael*, The *Angel of Brightness, Beauty, Healing, and Life*. Anyone who can "see" the human aura understands the relationship of these attributes. Your aura reveals the quality of your life and health. The brighter the aura the more beauty it displays. It measures the life force and your state of health. The brighter and more beautiful your aura the more vibrant your health.

The Realization of Conception for Tiphareth is *Beauty* and *equilibrium* in forms. Tiphareth is the *intermediary* between The Crown and The Kingdom. It is the *mediating* principle between Creator and creation.

As the intermediary between the Crown (Kether) and The

Kingdom (Malkuth), Tiphareth processes the energy of Pure White Light into energy forms that can be useful at the level of the material world. Energy intelligence is transformed into forces, laws, and reactions that govern the minute to minute activities and balances in the material plane.

The mediating principle runs both ways between the Creator and the creation. The energy information and ideas that are transmitted from the Creator to his creation are made available by Tiphareth as explained above. The reactions, ideas, beliefs, and desires from the level of the creation are all translated at the level of Tiphareth back into a light energy that the Creator can understand. This is explained in Romans 8:26, **"Likewise, the Spirit also helps our infirmities, for we know not what to pray for, as we ought. But the Spirit itself makes intercession for us with groanings which cannot be uttered."**

We have spent a lot of time learning about Tiphareth because it is at the center of everything. Tiphareth symbolizes many different ideas concerning the forming and perfecting of the Universe in which we live as well as symbolizing the ultimate purpose and destiny for Mankind. Divine Thought planned out the Universe and Man's place in it quite well, and watches it unfold according to His cosmic timetable.

Tiphareth reveals a Messiah Consciousness that was put into action from the very beginning of Universe. Emanating the Universe for our benefit and use was a monumental cosmic sacrifice by Divine Thought. The cosmic sacrifice was necessary on the material plane as well to put the Universe and mankind back into balance. We live in a day and age when people do not like to acknowledge that the sacrifice of Yahshua the Messiah was necessary. But it was necessary, not only for the restoration of the universe, but for the restoration of each individual as well. This sacrifice is called The Great Work.

The Great Work continues to operate in the Universe as well as within the individual human being who acknowledges it. Each of us has a part to play in The Great Work.

Now is the time for you to begin to do . . .

your part in The Great Work.

Tiphareth #6

1. Tiphareth is the seat of beauty, harmony and equilibrium.
2. All general health problems, and healings, take place at this level of consciousness.
3. Tiphareth *mediates* between all other levels of consciousness or Sephiroth.
4. The *virtue* of Tiphareth is *devotion to the Great Work.*
5. The *vice of* Tiphareth is *false pride.*
6. Tiphareth is known by several titles, some of which are:

   - *The Lesser Countenance*—a reflection of the unity of Kether through harmony.
   - *The King*—represents the creative substance (cornerstone) out of which the Universe was emanated.
   - *The Man*—bridges the gap between Heaven (Endless World) and Earth (material world).
   - *Adam*—represents Adam Kadmon, the spiritual pattern for material man.
   - *The Son*—refers to the Logos, The Word, The Only Begotten, and The First Begotten, not the "first created."
   - *The Sun*—symbolizes God imparting the life-giving energies that sustain the Earth and all of life upon it.

7. Tiphareth is also represented by *symbols,* some of which are:

   - *Symbol of a Child*—indicating that a child would be born to provide a sacrifice for the benefit of humanity.
   - *Symbol of a sacrificed God*—which reflects carrying out *The Great Work.*

8. *Yahshua* fulfilled the *titles* and the *symbols* of Tiphareth.

- *He* was *the lesser Countenance* or reflection of YAHWEH in that He claimed that He and the Father were One.
- *He* was called *the King*, the Cornerstone, and the Light, referring to the substance by which and out of which the Universe was emanated.
- *He* called himself the Son of *Man*, to symbolize bridging the gap between Heaven and Earth, or mediating between God and Man.
- *He* was called the Second *Adam*, or the Spiritual Adam to signify that He is the Adam Kadmon.
- *He* was called *The Son* of God to signify that He was the Logos, The Word, the First Begotten, and the Only Begotten Son. Only Begotten Son did *not* refer to his physical virgin birth, but to His prior emanation as the Logos.
- *He* was born as *a child*, was called "the Son of the Highest," and was given the name YAHOSHUA, which means "YAHWEH's salvation."
- *He* was the *Sacrificed God* who offered up his physical life for the benefit of Humanity as His part in doing *The Great Work*.

9. *The Great Work* for us, as Enlightened human beings, is to "present our bodies as a *living sacrifice* dedicated to *Divine Thought* for the benefit of a struggling Humanity.

Chapter 10

GEBURAH #5 AND CHESED #4

Geburah and Chesed will be studied together because it is only through the balance and equilibrium of these two forces that the Universe is sustained. Geburah is the Sephira of Justice, Judgment or Severity. It is on the Left Pillar of the Tree of Life, the Pillar of Judgment. Chesed is the Sephira of Mercy, Benevolence, or Love. It is on the Right Pillar, which is the Pillar of Mercy. *Justice and Mercy* are the highest attributes that a human being can achieve in relationship with others. This is one of the reasons that the Left Pillar and the Right Pillar are named after these two Sephiroth, or levels of consciousness.

The Old Testament of the Bible is full of examples of what people today would call strict laws and harsh judgments. It is not time to explain why that was necessary, but Yahweh's real desire for His people is stated twice in Hosea. After explaining that the whole nation of people were lying, stealing, cursing and killing each other, He stated in Hosea 6:6, **"For I desired *mercy* and not sacrifice, and the knowledge of YAHWAH more than burnt offerings."**

You see, when the knowledge of God is missing so are mercy and judgment. All you have left, without the knowledge of God are the selfish desires of human nature. Mercy and Judgment are not inherent in human nature. They are the highest level of energies flowing throughout the Universe and must be acquired by the human consciousness. That is why Yahweh also said in Hosea 12:6, **"Therefore turn to your God; keep *mercy and judgment*, and wait on your God continually."**

When you seek the higher levels of consciousness and you open yourself up to receiving the highest of spiritual attributes, you will begin to relate to others differently. Judgment and Mercy are not easy to understand. We all feel like we know when judgment has been too harsh, or too lenient. We can also think of instances where we feel that someone has displayed too much or too little mercy. But to understand the correct equilibrium of Judgment and Mercy requires that we be in proper balance. It requires that we be centered in Tiphareth.

In today's politically correct world people are being taught to be "non-judgmental." This is different than learning how to *balance* Judgment and Mercy. Judgment has a lot of different meanings and one of them is "the forming of an opinion or conclusion as from circumstances presented to the mind." Being non-judgmental is withholding your opinion or conclusion for fear of offending someone. More often than not, your opinion about someone else occupies a good portion of your thinking time and energy. In many people this can be a source of *dis-ease* because of the emotional drain on the energy system of their body. Judging or condemning another person is usually an attempt to control the attitudes and behavior of the other person. When the judgments are ignored or rejected, anger, disappointment or resentment set in and the result is a big time energy drain. Learning to be non-judgmental allows one to re-energize and re-center oneself.

The Apostle Paul in Romans 2:1-3 said it like this, **"you are inexcusable, O man, whomever you are that judges (condemns), for wherein you judge another you condemn yourself, because you do the same things."**

You see, most human beings spend so much time condemning and criticizing others that they do not see that they are guilty of the same things, even if only in their thoughts. Every thought we have has consequences. Every choice, and every feeling or emotion has personal, social, and even global consequences. This is why Yahshua the Messiah said in Matthew 12:36, **"I say unto you that every *idle word* that men shall speak they shall give account thereof in the Day of Judgment."**

*   *   *   *   *   *

I would like to stop you here for a moment and ask you to evaluate yourself with this question: "Do I have an energy drain on my power supply?"

You have already learned that your thoughts, feelings and emotions are various forms of energy. Everything you think about is a transfer of energy. The energy can be flowing to you, or from you. If your thoughts, feelings and emotions are positive, energy flows toward you and you feel energized! However, if the thoughts, feelings, and emotions are negative, then you will feel an energy drain. It is this constant ongoing energy drain that is the root cause of disease, illness and a lack of power or will. Every thought form to which you are attached is an energy drain. Every person or object of your thoughts, emotions and feelings is a thought form to which you are attached. Each thought form, to which you are negatively attached, needs to be detached from your personal power supply. You must *choose* to stop your negative feelings, emotions, and thoughts toward that thought form. By consciously pulling the plug on each offending thought form, you will begin to build up your energy, strength, and power of will. This is the benefit of becoming more non-judgmental toward others. At that point you will be able to see more clearly the balance between Judgment and Mercy.

*   *   *   *   *   *

What are Judgment and Mercy?
Judgment and Mercy are the result of several Emanations

that occurred before our present Universe was stabilized. The Hopi Indians as well as several other tribes have an account in their histories of several worlds that existed and were destroyed before our present world came to be. The ancient religions of the Eastern world have similar accounts in *their* histories. The Kabbalah states that there were 7 Worlds, or Universes that existed in succession before the present Universe was stabilized. The ultimate purpose of these previous worlds, of course, was to allow for all of the complexities that would be necessary to sustain Man and Man's purpose in this Universe. It was tantamount to creating a workable computer program that would give Man free will so as to accomplish his mission in life. Man must learn to fulfill his mission without destroying the Universe, and at the same time have enough latitude for reasoning and decision making to be able to affect intricate natural balances in the Universe.

The concept of Nirvana brings with it the idea of one undivided totality. Our ultimate purpose, from the Buddhist and Hindu viewpoints, is to become re-absorbed into that total Oneness, by shedding all imperfections of Individuality. This idea can be used to describe Divine Thought, The Ein Soph, and The Endless World. This is also the perfect Oneness of Adam Kadmon as it was first Emanated. Adam Kadmon was originally total Oneness, neither positive nor negative, male or female. But this presented a problem because no change was possible. The purpose of the Universe, after all, was to allow for the necessity of change through free Will, and the expression of Individuality.

No change or Individuality was possible in Adam Kadmon because when the 10 Sephiroth were originally emanated they were in a configuration of a straight line one under the other without any way to achieve a balance. No allowance was made for any kind of deviation from the perfect Oneness. The laws that regulated the Universe were unbending and unchallenged. There was no leniency.

Judgment was Absolute. Therefore that Universe self-destructed. Individuality became possible only after Adam

Kadmon was rendered male and female. The Tree of Life became reconfigured into its present state of two opposite Pillars balanced by the Middle Pillar of Equilibrium. This provided permanent stability. Everything in the Universe is governed by this Male-Female Principle. This is why the Kabbalah says that the world was first created by Judgment and could not subsist until Mercy was conjoined with Judgment so that Divine Mercies would sustain the Universe.

Geburah is strength, might, and severity, which is a feminine potency. This may seem like a contradiction but it is really a form of balance. Feminine means a receptive vessel. So to choose what will be received and what will not be received, requires a position of strength. That is why strength is a feminine force or energy. Justice rules through and from fear, which makes it the most forceful and disciplined of all of the Sephiroth. The essence of the energies of Geburah range from justice to abject cruelty, which makes Mercy a necessary balance.

The *healing energies* of Geburah pertain to the strength of our levels of energy. Anemia originates at this level as well as the resulting sinus problems, colds, fevers and body temperature fluctuations. Red corpuscles of the blood and the entire muscular system can also be influenced here. And this is where everything associated with the adrenal glands originates. Meditating at the level of Geburah can help you to unlock energy flows and increase your levels of energy and stamina.

Geburah can unleash greater power in your life. Power is the force or momentum of energy that gives you the ability to do or act. Power, like all other energies previously discussed, is intelligent energy. Greater power will lead to greater courage, initiative, and judgment. Courage is the quality of mind or of spirit to help you face a difficult situation, an immediate danger or severe pain. Initiative is the readiness to begin a new project or to make a change in an old project. Judgment is the ability to make decisions based on the facts or evidence at hand. These are all manifestations of the intelligent energy of Geburah.

It was stated in the beginning of this book that we are here on this earth to make changes in our spiritual lives. The reason you have read this far into the book is that you are searching for ways to make corrections in *your* life. You desire to know what alterations can be made and how to do it. The energies of Geburah will give you the *power* to make those desired adjustments. But before you can make any changes in your life, you must first alter your old thought patterns and thinking habits, and replace them with the new. As you have already learned, in order to manifest your thoughts they must be driven by emotion. The more powerful your emotions the more quickly will be the manifestation of your thoughts and desires. It is the strength and power of Geburah that empowers the emotions behind your thought patterns, whether old or newly replaced.

As we increase in power, it becomes necessary to learn to use and to control that power intelligently. The responsible use of power requires the use of critical judgment. Critical judgment is establishing and maintaining the highest principle on earth, Justice. There is no higher principle. Its force is not an evil force unless its essence overflows into cruelty. But when Judgment is rendered, it is based on the very energy intelligence that rules the Universe.

What is Justice?

Justice is the essential distinction between *good* and *evil*. It is the first truth of morality. Justice is not a consequence of anything. In other words it is not the result of any other action. For example, suppose a defendant is standing before a Judge in a Court of Law. The defendant is accused of stealing a car and for that reason is called to judgment. The Judge will issue a "Judgment" against the defendant based on the evidence provided. The defendant may be declared guilty or not guilty, and if found to be guilty will be required by the "Judgment" to make some kind of restitution. Based on the terms of the "Judgment" people will know whether or not *Justice* was upheld.

The highest principle of the Universe is to "receive for the purpose of giving to others." Every transaction in the Universe is based on the Primary Principle of "giving and receiving." Justice requires that both giving and receiving must be voluntary. This is a requirement of free will. A man who rapes a woman is "giving" to her something that she does not desire to "receive." That man has violated the highest principle of the Universe and therefore, *Justice* will be required of him.

In the court of Law example given above, whoever stole the car also violated the highest principle of the Universe. The car thief "received" something that the car owner did not desire to "give" or "to share." Justice for that person will be required, either now or Cosmically. In this case, the defendant will have received Justice if he/she were the actual thief and was found to be guilty. On the other hand, if the defendant were judged to be guilty, and was not the actual car thief, then Justice would have been miscarried, and Cosmic Adjustments would eventually be made. Justice can never be escaped if you live in the present Universe. This is called *Kharma.*

St. Thomas Aquinas once stated, **"A thing is not *just* because God wills it, but God wills it because it is *just.*"** *Ruach* is the part of the soul with which you most closely identify. Ruach is your Ego, or that which you think of when you identify yourself. Ruach is based on the interplay of 5 Sephiroth, those being Hod, Netzach, Tiphareth, Geburah and Chesed. In relation to Ruach these manifest respectively as *Reason*, *Desire*, *Imagination*, *Will* and *Memory*. It is the synthesis of these energies that determine whether you will be driven by the demands of your human nature of *Nephesh*, or will allow the Enlightenment of *Neshamah* to brighten your day.

Will is the Principle of Geburah, which is the *power* of the spiritual Self in action. The will can be your Servant, but most often it is your Master. It is the power by which you are enslaved to your passions, emotions, feelings, thought patterns, habits and beliefs. These are your Masters. They control your will. How can *you* gain control over your own Will?

There are *two steps* to gaining control over your own will. The first step is Forgiveness. You must forgive all whom you think have wronged you. Yahshua said in Matthew 6:12,14-15, **"Forgive** us our debts as we forgive our debtors . . . for if ye forgive men their debts, your Heavenly Father will also forgive you. But if you forgive not men (and women) their debts, neither will your Father forgive your debts."

You see, it is important for our own well being that we forgive others. Once we stop judging and condemning others and become more non-judgmental, we will find it easier to grant forgiveness to others. True forgiveness enables us to begin to assume more control over our own will.

The second step to gaining control over your will is to surrender your personal will to Divine Will. This may seem like a contradiction but it actually is not. You see there is a natural fear to surrendering your own will to the Will of Divine Thought. That fear is actually a part of Geburah. It is said, *"the world was created by judgment, which is fear."* Judgment is balanced by Mercy, which is Love. And the Bible says, *"Perfect love casts out fear."* Once you have surrendered to Divine Will you will discover a peace of mind that you have never experienced before.

It is said that Geburah and Chesed represent the expansive and contracting forces of the Universe. They also represent the forces of repulsion and attraction, and the magnetic force between the two dimensional poles, acting under the will of the Logos. Without Justice and Mercy working together the Universe could not be sustained. The inflexibility of *Justice* would not allow for any leniency in adaptability, fluctuation or changes. The first mistake, error, or sin would wipe out the entire Universe. On the other hand the strong arm of Justice prevents a world entirely of *Mercy* from degenerating into foolishness, cowardice and total chaos. It is the equal interaction between the two extremes that provides for the permanence and stability of the Universe. This allows for the adaptation and evolution of Nature upon the earth, and for the exercise of free will for spiritual advancement of the human soul.

The *virtue* of Geburah is *energy* and *courage.* Energy is the capacity for vigorous activity and for forcefulness of expression. Energy is necessary to accomplish the activities to fulfill your mission in life. Courage is the will to act in accordance with your own beliefs inspite of criticism from others. Energy and Courage require the power of will, and there is a source of intelligent energy that you can access to increase and strengthen your power of will. The Apostle Paul tells us in Ephesians 3:14-20, **"[Acknowledging] the Father of Yahshua the Messiah, of whom the whole family in heaven and earth is named, that he would grant you according to the riches of his glory, to be strengthened with power by his Spirit in the inner man. That Messiah may dwell in your hearts by faith; that you, being rooted and grounded in love, may be able to comprehend with all saints what is the breadth, and length, and depth and height; and to know the love of Messiah which passes knowledge, that you might be filled with all the fullness of God [Divine Thought]."**

The Rabbi Gamaliel taught the Apostle Paul the Kabbalah and the Tree of Life before Yahshua the Messiah taught him to deliver the message of the *"Abrahamic Kabbalah."* In the above quoted paragraph, Paul was explaining how to expand your power of will. The Father [Chockmah] would send the power through the Spirit (Mother) [Binah], to strengthen your inner man (soul) so that the Messiah may dwell in your heart [Tiphareth] so that if you are rooted and grounded in love [Chesed], you would begin to experience a broad range of energy intelligence that is beyond mere knowledge. This experience can give you power and courage [Geburah] way beyond what you would normally expect.

The *vice* of Geburah is *cruelty* and *destruction*. If not grounded in mercy and love, the desire for power can leave you cold, unfeeling, and abusive. You have heard the saying that *"Power corrupts, and absolute power corrupts absolutely."* When one becomes corrupted by power he becomes seized with the desire to take from others without giving anything in return. This leads to cruelty and eventual destruction. When seeking power it is very important to use that power for the benefit of others with their willing consent. To do so requires a balanced approach to the use of power.

The *Abrahamic Kabbalah* teaches that there are two sources of power available to mankind. The first source of power is the power (spirit) of the Cosmos, or Universal Power, which is taught by the traditional Kabbalah. The second source of power is taught by the oral traditions of the *Abrahamic Kabbalah*. Paul makes this distinction plain in I Corinthians 2:12-16, **"Now we have received not the spirit of the world, but the spirit, which is of God; that we might know the things that are freely given to us of God. Which things also we speak not in the words which man's wisdom teaches, but which the Holy Spirit teaches; comparing spiritual things with spiritual. But the natural man receives not the things of the Spirit of God; for they are foolishness to him; neither can he know them because they are spiritually discerned. But he that is spiritual judges (scrutinizes) all things, yet he himself is judged (scrutinized) of no man. For who has known the mind of the Master that he may instruct him? But we have the mind of Messiah."**

The energy available to the natural man, or general public, is based on the thought forms that have been the basis of organization and power for thousands of years, including the secrets of some of the Mystery Religions. These ancient thought forms are the basis for the corruption of power because they lack the proper balance of justice and mercy. This is called the spirit of the world.

There is exceptional energy available to those who are *"in the mind of Messiah* [Logos]." It is based on the thought forms pertaining to Yahshua the Messiah. This energy enables you to avoid descending into cruelty and destruction as you interact with others. Why? It is because Yahshua the Messiah is centered in Tiphareth and balances all forms and energies.

The *Archangel* pertaining to Geburah is *Khamel*. He is the Prince of *Strength* and *Courage*. He protects the weak and defends the wronged against their accusers. The Angelic Order at this level is the Seraphim (fiery serpents) who assist us in stopping those who would overthrow and upset our world order, and our individual lives. There have been several accounts in recent history that white shining beings riding glistening white horses have intervened to turn the course of a crucial battle. One example happened during World War II. A German advance was just about to begin that would overrun an allied position that was in sore need of reinforcement. A German victory here could have changed the course of history. Suddenly troops on both sides saw a Cavalry of larger than life Beings armed with swords, riding white horses between the opposing lines. Allied Soldiers asked each other, who or what was that? German Officers were fearful and delayed beginning their advance to await an explanation and instructions from Headquarters. In the meantime allied reinforcements arrived and they were able to fortify and hold their position. Soldiers and officers of both sides of the war have since been interviewed about their memories of that unforgettable event. Similar situations have taken place at other battlefronts in other wars as well, most recently was during the Arab-Israeli 6-Day War in 1967.

This has also happened on an individual level for many people. Often this protection is totally unseen, but not always. New Tribes Missionaries are taught to accomplish their Mission or Great Work completely on faith. They are sent to tribes of people in Africa or South America who have never received the message about A Savior who provided salvation for them. Each Missionary, having no money, goes to the Airport expecting "God to provide" for them. Eventually someone inquires, and after a conversation provides them money or purchases a ticket for them. One man in particular finally arrived at his destination only to find the tribe of people very hostile to his message. One day they took Council and determined to kill him that night. When they descended on the man's hut as he slept unaware, they suddenly stopped and looked at each other in terror. They saw a squad of Beings armed with swords surrounding the hut facing them. Silently they crept back home. Next day they returned to the man's hut and explained that they were ready to listen to his message. They explained the entire episode of the night before and believed that if he had that kind of protection his message must be important.

Matters were no different thousands of years ago than they are today, because a similar event is recorded in the Bible concerning Elisha the Prophet. It began with the king of Syria desiring to destroy the army of Israel so they could rule the land of Israel. Each day he would lay a trap for the Army of Israel, but Elisha, through Clairaudience, was able to hear the plans of the King of Syria, and to warn the King of Israel to avoid each of those traps. Eventually, the King of Syria discovered that Elisha was the problem, so he sent his army out to capture Elisha. Elisha, knowing of the plans left the city early and went out to the mountains to meet the army. The servant of Elisha said, "Oh no, what are we going to do?" Elisha responded in II Kings 6:17, **"Lord, I pray thee open his eyes that he may see. And the Lord opened the eyes of the young man and he saw; and behold, the mountain was full of horses and chariots of fire round about Elisha."**

Elisha was protected and he was able to control the situation to where the King of Syria abandoned any further plans of attack on Israel. You see, there is a Great Work being accomplished in the material plane and the Angels of Justice ensure that the Work continues on its course.

The name of God at this level is *Elohim Gebor*, meaning, *God Almighty*. He is the source of All Power, which manifests from this level to avenge and to judge evil. *God of Battle* is another title of this level, and it is said that his steps are Lightning and Flame. This indicates that Power is used at this level to burn away, or remove all that is useless, so that the "New" may emerge.

This same idea is revealed in I Peter 5:8-10, **"Be sober, be vigilant; because your adversary the devil, as a roaring lion, walks about seeking whom he may devour. Whom resist steadfast in the faith, knowing that the same afflictions are** accomplished in your brothers, that are in the world. But the God of all grace, who has called us unto his eternal glory by the Messiah Yahshua, *after you have suffered awhile*, make you perfect, stablish, strengthen, settle you."

Your problems are being personified as being brought on by "the devil," but whatever the source of your suffering, its purpose is to make you perfect (complete), strong and established. Suffering is the fire that burns away the old, useless thought patterns, feelings and emotions, so they can be replaced with *thinking patterns* that are stronger and more stable. With this new strength you can overcome any evil.

The Realization of Conception in Symbolic Form is *Severity* necessitated by Wisdom and good will. Severity is necessary in order to establish and to protect the good. To permit evil to abound is to hinder good, and for this reason Judgment came into the world. John 9:39, **"And Yahshua said, for judgment I am come into this world, that they which see *not* might see, and they which see might be blind."**

Yahshua came to reveal the true balance between Judgment and Mercy so that the spiritually blind might see. On the other hand those that "see" and allow evil to abound will be judged for their actions.

Chesed emanated from Binah and is a masculine potency. Chesed is located on the Pillar of Mercy directly across from Geburah. Chesed means Mercy and Love. This is the highest Sephira, or level of consciousness that can be conceived by the human mind. This is why Love, or Mercy, has been called the most powerful force in the Universe.

Chesed is also called Gedulah, which means Greatness and Magnificence. It is a reflection of Chockmah, which is known as the All Knowing Father. Gedulah, or Chesed, reflects the Father as the loving, forgiving, and protecting Father, who was revealed to us by the Messiah, Yahshua. Matthew 11:27, **"All things are delivered unto me of my Father; and no one knows the Son but the Father; neither does anyone know the Father but the Son and he or she to whom the Son will reveal."**

In the ancient world the All Knowing Father was known by his attributes which related to fear, power, war and judgment. It had been forgotten that YAHWEH wanted justice and mercy from his people, and not sacrifice. It was Yahshua who came and revealed the loving, forgiving, and protecting Father. He revealed The Father by the highest attribute that the human mind could conceive, which is *Mercy* and *Love*, and most people, even today, still do not comprehend it.

Areas of health that can be affected at this level are formation of hemoglobin, cell nutrition, intestines, liver, hips, thighs, mouth, throat, neck, thyroid, parathyroid, and hypothalamus.

Chesed is where we can manifest greater *abundance* in our lives. This is the level of growth and prosperity and is where good organization can produce the greatest productivity. This is where you become most effective in doing The Great Work.

Chesed is also the level of peace and mercy, and the exercise of true power at the highest spiritual levels. Chesed is the consciousness level at which you learn to use the power of *Love* without judgment, assessment or other conditions. This is described in I Corinthians 13:2-7, **Though I have the gift of prophecy and understand all mysteries, and all knowledge, and though I had all faith so that I could move mountains, and have not love I am nothing. And though I give all my goods to feed the poor and though I give my body to be burned, and have not love, it profits me nothing. Love suffers long and is kind. Love envies not; love boasts not itself, is not puffed up; does not behave itself unseemly, seeks not her own, is not easily provoked, thinks no evil. Love rejoices not in iniquity but rejoices in the truth. Love bears all things, believes all things, hopes all things, endures all things."**

However, love without judgment is not unchallenged Mercy. Justice is a quality of energy in the Universe that is immutable and unchanging. We know that justice is handed down to us throughout our life as Kharma, and we know that justice awaits us after death when we are judged to ascertain how we have conducted our life plan. Justice is a balance of Judgment and Mercy. It is commonly assumed that The Father is our final Judge, but this is only half of the story. The rest of the story is found in John 5:21-23, 26,27,30 **"For as the Father raises up the dead and brings them back to life, even so the Son brings back to life whom he will. For the Father judges no one but has committed all judgment unto the Son; that all men should honor the Son even as they honor the Father. He that honors not the Son honors not the Father who has sent him . . . For as the Father has life in himself so has he given to the Son to have life in himself; and has given him authority to execute judgment also, because he is the Son of man . . . I can of my own self do nothing; as I hear I judge. My judgment is just, because I seek not my own will but the will of the Father who has sent me."**

YAHSHUA was born into this world to provide a "vehicle" for Divine Thought to set the balance for Judgment and Mercy. This is why YAHSHUA was given the power and authority of Judgment over the affairs of all Mankind. And because the final verdict is rendered for each and every individual after death, Yahshua was also given the power over life and death. Today, it is politically incorrect to "honor the Son" or to even admit that Yahshua has the power of eternal life over each and every soul that has walked upon the earth. But it is true, never the less.

The *Virtue* of Chesed is *obedience to a higher will*. Yahshua gave total obedience to Divine Will. His personal will and Divine Will were merged as one. When we enter this level of consciousness we become ready and willing to merge our own will with Divine Will. When this happens the physical path and the spiritual path lead to the same destination. You become at peace and begin to take genuine pleasure in life. And then you become ready to be liberated from binding physical illusions.

There are people who think it a terrible thing for Divine Thought to "sacrifice his Son" by crucifixion and death for the salvation of humanity. In reality the sacrifice was totally voluntary. Yahshua *chose* to accept his mission and complete The Great Work. He said in John 10:17-18, **"Therefore does my Father love me because I lay down my life that I might take it again. No one takes it from me, but I lay it down myself. I have power to lay it down and I have power to take it again. This commandment have I received of my Father."**

Everything works out for the best when your will is in obedience to Divine Will. In Yahshua's case he knew it would be like a near death experience, because after 3 days he would reclaim his life, body and soul united forever. The same can be true for us if we are willing to unite our personal will to the will of the Father.

The *vice* of Chesed is *hypocrisy* and *bigotry*. Hypocrisy is the result of the *fear* of surrendering you personal will to Divine Will. Hypocrisy is the art of pretending to be doing the Will of the Father as an outward show to others but showing no fruit of performing the Divine Will. Bigotry is judgment against someone based on various forms of material or physical attributes, while ignoring the spiritual qualities. All of this is based on fear of Divine Will.

The *Archangel* ruling this plane is *Tzadkiel*, the Prince of Mercy and Beneficence. His name means *Righteousness of God*. It is said that he was the protecting angel of Abraham. The class of angels here is the Chasmalim, the Brilliant Ones. They manifest the brilliance, power and majesty of Yahweh.

The manifestation of God in Chesed is *EL, The Mighty One.* He rules with glory, magnificence and grace. I Timothy 3:16 reveals, **"Without controversy great is the mystery of godliness. God was manifest in the flesh, justified in the Spirit, seen of Angels, preached unto the gentiles, believed on in the world, received up into glory."**

Divine Thought can manifest itself within any level of consciousness. This is why people at any level of consciousness can have a mystical experience. For some it my be an experience of being enveloped in white light that endows a sense of peace and enlightenment. Others may see a vision of a Being that they perceive as God who communicates a message to them. Still others may receive audible messages in their head. Divine Thought can also manifest on the material plane in the flesh, as was the case when he appeared to Abraham and sat down and ate with him before going over to execute judgment on Sodom and Gomorrah.

In order to complete the first phase of The Great Work it was necessary for Divine Thought to experience death. This is why Divine Thought, as the Logos or The Word, was inserted into the body of Yahshua. In this way, everything human could be experienced. By achieving Victory over Death while dwelling in the flesh, it was made possible for us to do the same. Today, *El* rules with magnificence, glory and grace within the minds and hearts of those who believe. This is known as the mystery of godliness.

The Realization of Conception in Symbolic Form is *Mercy*, which is Wisdom in its secondary conception, strong and benevolent. Mercy is a reflection of Wisdom, which cannot be understood by the natural human mind. Mercy is the highest quality of spiritual energy that man's mind *can* conceive. But mercy for the sake of mercy is useless. Mercy must be strong and benevolent so that it is used for a positive purpose. Extending mercy to someone should only be for the purpose of advancing his or her spiritual journey. Otherwise justice goes unserved and the mercy is undeserved which adds to the kharma of both that person and *you*. You see, when justice on earth is not met then *Cosmic Justice* makes adjustments.

In summary, this chapter on Judgment and Mercy is an introduction to the highest attributes and spiritual qualities that we as children of Divine Thought can manifest in our own lives toward others. In fact, this chapter depicts our part in *The Great Work*, which we spend much of our lives trying to recognize. Our purpose in this life is to learn how to properly use Judgment and Mercy in all of our relationships with others, whether family, friends, business associates, or strangers. Now is the time to find an endeavor through which we may learn how to balance and administer the spiritual qualities of . . .

Judgment and Mercy!

Geburah #5, and Chesed #4

1. Geburah is the consciousness level of Judgment, or severity.
2. Divine Thought expects us to practice proper Judgment and Mercy.
3. An attitude of judgment and condemnation of others is harmful to oneself. It is an energy drain.
4. Health problems here relate to low energy levels, anemia, and stamina.
5. The Universe could not exist without a balance of Judgment and Mercy.
6. Judgment is too restrictive and does not allow for change or adaptation.
7. *Ruach,* your Ego, is the mental manifestation of your soul. Geburah adds willpower to your Ego or soul.
8. Surrendering to *Divine Will* brings tranquility and peace of mind to your soul.
9. The *virtue* of Geburah is *energy* and *courage.*
10. The *vice* of Geburah is *cruelty* and *destruction.*
11. Chesed means Mercy and Love.
12. Mercy and love are the highest energy intelligences available to man's mind.
13. The Universe could not exist in total Mercy without Judgment. It would be unbridled chaos.
14. The *virtue* of Chesed is *obedience to a higher will.*
15. The *vice* of Chesed is *hypocrisy* and *bigotry.*
16. *Yahshua the Messiah* administers Judgment and Mercy on the earth and in the world to come.

Chapter 11

THE SUPERNALS

The first 3 Sephiroth of the Tree of Life are called the *Supernals*. These three provide the substance out of which the lower seven are emanated. The lower 7, which we have just finished studying, make up the energy levels that can be sensed by the human mind. The energy levels of the Supernals are not reachable by the human mind. The Supernals make up the levels of existence that for lack of a better word, are called God, or *Divine Thought*.

The lower 7 Sephiroth, or levels of consciousness, are energy levels that are available to human beings even if they do not believe in a God. They are attributes of God, or Divine Thought, that can be emulated by humanity in an effort to develop the most effective relationships with one another. Even though human beings may claim these qualities and attributes as their own, they still emanate to them from a higher, undetected source, which is the Supernals.

The Supernals are separated from the lower 7 by a gulf known as *The Abyss*. This is a barrier that separates Divine Thought from the various energy levels of the material world and the thought processes of the human mind. The material Universe and all of its levels that we have available to us, is a manifestation of an Ideal pattern that has always existed in the mind of Divine

Thought. When the Bible says, **"In the beginning God created the heavens and the earth . . . "** it is speaking of the beginning of the manifestation of the material Universe. The Universe always existed in potential form but its manifestation began as it crossed the great Abyss. This phenomenon is explained in Genesis 1:2-3, **"And the earth became without form and void, and darkness was upon the face of the deep. And the spirit of God moved upon the face of the waters. And God said, let there be light and there was light."**

The manifestation of the *substance* out of which the Universe sprang began when God said, "Let there be Light." And from that substance that the Kabbalah calls *White Light*, sprang the Universe and all of the laws and subtleties that govern it. But you will notice that it was the action of the spirit of God moving across the *Abyss* (darkness, void, formlessness, the deep, and the waters) that began the Emanation process.

One of the discoveries from the latest telescope in earth orbit is that the Scientists involved believe they may have photographed the outer edges of the Universe. What they may have seen looks like a foamy substance beyond which they cannot see. If this is true, then it could be the barrier between the spiritual and the material world, or the *Abyss*.

When a person dies the departing soul usually sees a point of light and is drawn to it. The soul then approaches, or enters a corridor of bright light where it is determined whether or not this is the correct time for that soul to move on to the next world, or return back to this one. Once the soul moves through the corridor of light, which crosses the Abyss, it will not normally return to the material world except to reincarnate into the body of a newly born baby.

The Supernals constitute the head of Adam Kadmon, but they are just a reflection of the Endless World. *The Endless World is Divine Thought.* This is where the pattern for the material Universe is, and has always been. There has never been a time when there was not a pattern for the Universe and all that is in it. How many times has the Universe been manifested and then dissolved after billions of years of existence, we shall never know. But that pattern is Divine Thought. The Endless World in a state of repose or inactivity is called *Ain*, which means One or Indivisible.

All of the ancient religions had similar teachings about the Nature of God and the Origin of the Universe. This is how we know that Primitive Man was taught about God, the nature of the Universe, and man's place in it. The names have been changed but the stories are all the same. For example, Lao Tsu taught, **"Tao produced unity, unity produced duality, duality produced trinity, and trinity produced all existing things."**

Kabbalah teaches that *Ain* is the Unknown Causeless Cause. It is a state of existence to which we can never go or even know. It is the Unknowable. It is a state of pure negativity, the opposite of our world of positivity. The closest understanding of this in our material world would be the idea of antimatter. One of the theories of achieving faster than light travel is called warp speed. Even today the Huntsville Space Center is researching ideas relating to warping space. It takes a tremendous amount of energy to accomplish that task and nuclear energy cannot produce that magnitude of power. However, the energy contained in all of the levels of matter that govern it and hold it together, could be a source of tremendous power. Combining it with its equal opposite would unleash a tremendous amount of power. This is the theory of antimatter. But controlling antimatter would be like trying to capture and control a black hole the size of the period at the end of this sentence. Ain desires only one thing, and that is to bestow or impart "his" essence. But to whom will "he" impart?

Ain created a container, or Cosmic Womb to receive "his" essence. It might be likened to a *force field* that would contain and separate that which was inside from that which was outside. Ain was outside. The container contained negative energy but it desired something else. The desire to receive marked the beginning of the change from inactivity and repose to activity. This container was called *Ain Soph*. Kabbalah teaches that Ain Soph contains a limitless or infinite amount of energy. Ain Soph is the very beginning of self-consciousness because this is where unity becomes duality. Self-consciousness arises when there is the ability and awareness to make distinctions and comparisons. Ain Soph is also the origin of Self-Comprehension. Knowing who you are, what you are composed of, and the extent of your purpose all originated here. Ain Soph is the force field that determines the outer limits of our Universe. The desire of Ain Soph to receive was satisfied when Ain imparted to "her" his essence of Limitless White Light.

The union of the duality of Ain and Ain Soph produced the trinity, and completed the 3 planes of unmanifestation. The Absolute Limitless Light in Ain Soph is called *Ain Soph Aur*. The Limitless White Light carried within it the seeds of Positive Existence. Positive existence relates to the idea of the Archetypal Man or the Adam Kadmon. When manifested, the Adam Kadmon would be the body of Divine Thought.

When the essence of the positive White Light was imparted into the negative energy within Ain Soph, it was quickly condensed into a small point of light in the center of the container, Ain Soph. This central, dimensionless point is called Kether, which means Corona or Crown.

Sephiroth #1, Kether.

Kether is called the Ancient of Ancients and the Cause of Causes. It is the aggregate of all 10 Sephiroth, as potentialities, but indistinguishable while in Kether. Only when the energy emanates from Kether does each remaining Sephira take on its own characteristics. Kether is the pure *Will* of Divine Thought. The Will of Divine Thought is the Soul of all that exists. When Kether emanated from the Ain Soph, it manifested as both the vessel and the substance within the vessel.

It has been described in this manner. Imagine an ocean of light blue water, and centered within it a sphere of denser, darker, blue-green water. The outer surface of this sphere is the vessel, which contains the denser water. The vessel is composed of the same substance contained within, but is more like a force field. This is similar to what may have been observed by scientists at the outer edge of the Universe.

Divine Thought is the *Idea* of a Universe. The Idea of the Universe takes on human form and is called Adam Kadmon. Kether represents the cranium while Chockmah and Binah, the following two Sephiroth are the two lobes of the brain. Chesed and Geburah are the two arms, Tiphareth is the trunk, and Netzach and Hod are the thighs. Yesod is the female organ and Malkuth the male organ of reproduction. As the neck separates the head from the body, so does the Abyss separate the Supernals from the Lower 7. Kether is the beginning of all things and the maker of all things. It is the source of all ideas, whether past, present, or yet to come. Kether, the Corona, is somewhat like an embryo. It begins in a small, spherical shape but eventually expands into the shape and image of a human being with a head, arms, legs and torso. All of the other 9 Sephiroth are contained in Kether until they expand into the shape of the Tree of Life.

The *Sepher Yetsirah* calls Kether "the Hidden Intelligence." As such each and every Sephira possesses 3 distinct qualities. First of all, it possesses its own unique qualities, and level of consciousness, that it can offer. Secondly, it receives from the previous Sephira everything that it needs to maintain its own unique qualities. In the case of Kether, it receives from the Endless World, or Ain Soph. The third quality is that each Sephira transmits its own nature, and everything received from above, to the Sephiroth below. This being the case, there is but one indivisible and absolute consciousness permeating each and every particle and space in the manifested Universe. But its first differentiation gives rise to what men might call *gods*, which are actually levels of consciousness that the human mind cannot conceive. These *gods* are actually the *forces* that govern and maintain the structure and operation of the Universe and everything in it. They are known as the laws of nature, the force of gravity, the qualities of electromagnetics, the binding qualities of chemistry and physics, and the instinctive intelligences possessed by every specie in general, and each individual living creatures in particular. These "gods" are omnipotent, omnipresent, and eternal within the course of time wherein they are programmed or manifested.

It has been said that the names of the Gods are important because "the knowledge of the name gave power over its owner." If you are walking down a street and someone calls out to you, "Hey, come over here I want to talk to you."

Do you feel obligated to go and find out what they want? Not usually. Often you will completely ignore it, especially if it is accompanied with catcalls or whistles. However, if a stranger calls out *your name* and asks you to come over and talk, won't you feel more compelled to find out how they got your name and why?

So it is that knowing the name God, or an Archangel, or a class of angels at a given level of Consciousness gives you access to some of the attributes and qualities that the name controls or identifies. This is why each Sephira has a name of God, or description that identifies it.

The name of God at Kether is *Eheieh*, which means, "I will be." One day Moses was walking in the hills and he saw what looked like a bush that was on fire. He saw no smoke and the bush was not consumed, so he decided to walk over and take a closer look. As he cautiously moved in for a closer look, he was thinking, "Man, what in the world is this?" when suddenly a voice boomed out, **"stop! Come no closer! And take off your shoes because you are standing on sacred ground!"**

Moses was startled and shaken and quickly did as he was instructed. The voice continued to instruct Moses about leading the Children of Israel out of Egypt, but Moses was hesitant because he did not believe that the people would follow him. So Moses asked, **"When I say unto them the God of your fathers has sent me unto you, and they ask me 'what is his name?' what shall I answer?"**

The voice said, "*EHEIEH* [I AM THAT I AM]. Tell them, 'I AM (or I will be) has sent you.'"

Do you see what I AM means as a title of God? It means, "I exist." I AM THAT I AM then, means "I exist for the purpose of bringing into existence." That is also the purpose of Kether. Divine Thought wanted Moses to understand the *highest* possible manifestation of God from the Tree of Life. You see, Moses would eventually need to use the highest manifestation of the powers of Divine Thought to defeat the sorcerers and magicians of Egypt.

The *Archangel* is *Metatron*, which is actually the power behind the outflow of endless energy. This endless outflow of energy is what sustains mankind. The Kabbalah teaches that Divine Thought never *was not*, never *thought not*, and *never began* the creation of the Universe. The Archetype of the Universe never *did not exist* in Divine Thought. The spiritual essence always remains, so that when the Universe ends the Archetype continues to remain. Therefore, the Universe is dissolved and renewed in endless succession. When Man is in the Universe, Metatrone directs the endless outflow of energy toward him. It is said that Metatrone gave to Man the Kabbalah so that he might regain his true destiny.

The order of Angels at Kether is called *Chaioth ha-Qadesh*, The *Holy Living Creatures*. They are known as the angels of love, light, and fire. They are often called Seraphim. They can help us to better understand the Kabbalah.

One of the titles of Kether is *The First Swirlings*. It describes the meaning of Kether or Corona. This is an indication of the beginning of differentiation of cosmic energies. Kether is the doorstep into the Endless World. Ezekiel chapter One, describes seeing the visions of God. He says it began with a whirlwind engulfed in fire and the brightness of lights. Out of the brightness of the fire shot forth strikes of lightning, and in the center of it all were the 4 *Living Creatures*.

These symbols were all indications that the vision he saw came from the consciousness of Kether. The energy levels at Kether are beyond human grasp, but our higher soul, *Neshamah*, can put that energy information into a thought pattern that can be grasped. That is why Ezekiel could see strange sights and hear unusual sounds that extended beyond the normal human range. In this way Divine Thought could convey the idea that the forthcoming communications to Ezekiel were emanating from the very highest authority in the Universe. Calling him Son of man, Divine Thought then conveyed a message to Ezekiel, which would become his part in *The Great Work*.

Kether is considered the wellspring of Omnipotent, Divine Will. Beyond Kether there is nothing but Divine Will. But in our world Divine Will manifests as the Adam Kadmon, the ideal toward which we must aspire and achieve. As we advance upward in consciousness from Malkuth into the higher planes, we become conscious of Divine Thought, and the existence of Divine Will. Eventually we realize that our best interests are served by aligning ourselves with Divine Will. Our task then becomes learning to merge our own will with Divine Will. When we reach the level of Tiphareth, we are ready to devote ourselves to doing The Great Work. The Great Work is merging your will with Divine Will, and helping others to do the same.

In Kether there is only Unity. There is no duality. There is no differentiation. There is only One. Therefore, there is *no vice* in Kether. There is only virtue.

The *virtue* of Kether is *"Completion of the Great Work."*

The Great Work assigned to Ezekiel was for him to deliver a series of messages to Israel informing them that if they did not re-align themselves with Divine Will they would be driven by their enemies into exile. The Great Work for Israel was always for them to align themselves with Divine Will, and to teach the surrounding nations to do the same. Having failed to do so they would be scattered among the nations. However, after a period of exile, Israel would be gathered from among the nations and given another opportunity to resume their part in doing the Great Work. Today, Israel is a large nation, having been gathered from among the nations under another name, and Divine Thought is watching to see if she will begin to do her part in helping to complete The Great Work.

The Great Work for us, as individuals, is to discover Divine Will, and then begin the difficult process of merging our selfish will of human desires with Divine Will. Many people hesitate to even search for Divine Will because they believe that they will be required to try to negate their human desires. You will not be expected to negate your human desires, but to balance them as a lower priority to doing the Great Work. Most people do not achieve the Completion of the Great Work in their lives until the time of their death.

The realization of Conception in Symbolic Form for Kether is *The Crown*, which is the equilibrating power. As the energy emanates from Kether to the lower levels, the energy to always maintain balance accompanies it. For that reason, the Universe always remains in balance and equilibrium. And when an imbalance develops the forces are always there to begin to restore the equilibrium. That force continually emanates from Kether. On a personal level Kether is a source of greater creativity. On the other hand it is also the source of all information on the final ending or conclusion of any matter. Kether is where you can find answers to the causes of your inner spiritual quest and how to attain it.

Sephiroth #2, Chockmah

Kether is called the point, and Chockmah is the extension of the point into a line. This uplifted line is called the "rod of power." Chockmah is the first emanation from Kether and is the male, vigorous, active, vital, energizing element of existence. Chockmah is the beginning of the duality, out of which the Universe sprang. The male element is the generative power that activates Divine Will. It is the Divine intellectual power to generate thought. Chockmah is not thought, but is the generator that can initiate thought.

Chockmah is the consciousness level of Wisdom. It is the beginning of Potentiality. It is the very beginning of motion and action. Kether is the absolute Unity, or One. It is simultaneously the One and All. It is Divine Will which lays dormant and without action. Its emanation into Chockmah is also Unity, or One. But it is a reflection of the unity of Kether. Chockmah has the potential of departing from that Oneness into Duality, which is the beginning of action, motion, Intelligence, and differentiation from One into All. One is a Unity while All is an aggregate. As All, One and its reflection become Two. This is the level of Wisdom.

Wisdom is the beginning from which manifestation flows. When Wisdom is expanded from flowing forth, it is called "Father of Fathers." This is because Wisdom is the generative source of all that exists, and is the source of the beginning and the end of all things. For this reason Wisdom is called *Abba*, which means Father.

In **Mark 14:36,** When Yahshua was agonizing over completing his part in The Great Work, he said, **"Abba, Father, all things are possible for you. Take this cup from me. Nevertheless, not what I will, but what *you* will."**

Yahshua's human nature did not want to go through the pain and suffering of the crucifixion. But desiring to complete his part in The Great Work, he quickly realigned his personal will with Divine Will. But you will notice that his appeal was to Abba, The Father, the generative source of the Universe.

Chockmah is a level of Consciousness where we can obtain greater personal initiative. We can gain that source of energy, which puts things into motion, and helps us to gain the realization of our innate abilities. Chockmah is where we can gain or come to understand all Father type of information.

Chockmah has been called The Illuminating Intelligence. This is because it sheds light, or is a source of enlightenment. Chockmah is the Sephira of the Logos, or The Word. The Word was and is the creative force of the Universe, which initiated the beginning of the material emanation. Let us look at *John chapter one* in the Bible again, beginning in verse one: **"In the beginning was the Word, and the Word was with God (Kether), and the Word was God (Chockmah) . . . All things were made by him, and without him was not anything made that was made. In him was life, and the life was the light of men."**

As you can see, in Kether, God is Absolute Unity,

simultaneously One and All. In Chockmah, God is still Absolute Unity, and still simultaneously One and All, but with the added potential of action. The Word provided the action of emanating All things from the One. The Word provided all of the necessary ingredients for life. And the Word provided for the *Illuminating Intelligence* (light) of men.

You might ask, "Why are The Word and Abba, the Father, part of the same Sephira, if one is the son of the other?" In the Gospel of Thomas, Logion 61, Yahshua said, **"I am he who exists from the undivided."** Yahshua also said in *John 10:30,* **"I and my Father are one."** You see, Divine Thought in Kether is the potential for the Emanation of all things. But the reality of the Emanation actually began at Chockmah. When the essence of Chockmah Emanated to the next level Binah, then the Father came into existence. It is the essence, or activated "seed" by which Divine Thought became the Father. It is also the activated seed that Emanated the Universe. The activating principle was the spoken *Word.*

The healing energies of Chockmah influence the pituitary gland, the sinuses, and provide balance for the two hemispheres of the brain.

Chockmah is a Unity, which has the potential for diversity. By Emanating its equal opposite, Chockmah became part of a duality. But of itself it is a Unity. For this reason Chockmah has *no vice.*

The *Virtue* of Chockmah is *devotion.* That is the setting apart from all else to the single-minded purpose of someone or something. Chockmah is the first emanation and represents the number One. Kether represents "0" or infinity. Zero is undefinable, whereas "One" can be defined, according to the Kabbalist. Since all things and all numbers have a specific vibration, so does the number One. But it vibrates alternately from changelessness to definition and back to changelessness. Devotion illustrates this in the sense that an individual's sense of purpose is centered in the purpose of another, and the two purposes become one. Chockmah is the source of an individual's devotion to Divine Will. It is a very powerful vibrational energy. It is the result of Wisdom.

The manifestation of God at this level is *Yah,* or *Yahoweh*

(Jehovah). It means the Eternal, Self-existent one. This was the level of the manifestation of God that was introduced to the people of Israel, which was a much higher level of consciousness than the god levels of the surrounding nations, the Elohim. Apparently, Divine Thought expected more from Israel than from the neighboring nations.

The order of *angels* in Chockmah is *Auphanim*, or *Wheels*. They are also identified with the Zodiac. The energy-information identified with the Zodiac originates right here in Chockmah. It is part of the perfect Wisdom of Divine Will. The *Archangel* is *Ratziel*, Prince of Knowledge of the Hidden and Concealed. He can open up to us the Wisdom of the Universe and its operations.

The Realization of Conception in Symbolic Form for Chockmah is *Wisdom*, equilibrated in its unchangeable order by the initiative of intelligence. Everything originates or emanates from this level in a balanced and orderly manner.

Sephiroth #3, Binah

Binah is closely tied to Chockmah and is necessary to give Chockmah its own individuality. Binah means *Understanding*, which is the ability to grasp the concepts inherent in Wisdom. One cannot easily be defined without the other. For example, the Zohar calls Wisdom the Intellectual Generative Energy, and Understanding, the capacity to be impregnated by the active Generative Energy of Wisdom. The result produces Intellection, or Thought. It has been said that by Wisdom Divine Thought creates, and by Understanding, it establishes.

According to the Kabbalah, Kether is *Divine Will*, Chockmah is the *Intellectual power* to generate thought, and Binah is the *Intellectual capacity* to produce thought. However, they are not thought, nor do they contain the thought. Thought is the union of these two forces much like a zygote is the union of an egg and sperm. For that reason Binah is the feminine, negative, passive opposite of Chockmah. Binah is called The Fertile Mother, eternally conjoined with Abba, The Father for the maintenance of the order of the Universe. Binah is also called Marah (Mary), The Great Sea, and The Mother of All Living. Binah is the container, or Archetypal womb, which manifests all life. Therefore, Binah is the source of all Mother type of information. It has been said that The Father knows all and The Mother understands all.

Binah is the substantive vehicle of every possible phenomenon, whether physical or mental. Binah unites and brings together all forms and is the constructive power in the formation of all things. Binah actually carries out the plan of Divine Thought.

The healing energies of Binah affect the skin, bones, joints, tendons, teeth, spleen, hearing, colds and congestion in the body. So to influence the healing of any of these areas of the body go to this part of the Tree of Life.

The manifestation of *God* in Binah is *Yahoweh Elohim*. Binah is the feminine aspect of God, which is known as Elohim. Chockmah and Binah together are The Elohim. Yahoweh is the Eternal Power, and Elohim is the container or form through which that power is manifested and controlled. Elohim are both masculine and feminine so can refer either to gods or goddesses.

Yahoweh (Yahweh) Elohim manifests the perfection of creation, as in the proton and the electron, which are the building blocks of the material realm. Yahweh can be likened to the wise fountain eternally gushing with Wisdom, while Elohim is like the ocean or reservoir filled with Understanding. But the Wisdom cannot manifest of its own without the presence of Understanding. Binah is a reservoir of unlimited Potentiality.

The *Archangel* in Binah is *Tzaphkiel*, the Prince of Spiritual Strife Against Evil. He helps us to understand and overcome obstacles in a most spiritual and effective manner. He is also the Keeper of the *Akashic Records*, which can be revealed to us to help us to better understand the circumstances of our existence. The Akashic records are available to us at the level of Yesod, but are maintained for the use of Divine Thought at the level of Binah.

The order of Angels at this level is the *Aralim*, sometimes called *The Thrones*. The Throne is identified as the "seat of power" or the source from which power flows. Aralim are also referred to as "The Strong and Mighty Ones." They are available to help empower us through strife so we can achieve better understanding. This is hinted at in Colossians 1:9-11; " . . . [we] desire that you might be filled with the knowledge of his will in all *wisdom* and spiritual *understanding*, that you might walk worthy of the Master, unto all pleasing, being fruitful in every *good work*, and increasing in the knowledge of God, *strengthened* with all *might* according to his glorious *power* . . . "

Wisdom and Understanding are energy levels that are impossible to comprehend unless you actually have it. That is why King Solomon wrote in Proverbs 3:13-19, "Happy is the man that finds wisdom and the man that gets understanding. For the merchandise of it is better than the merchandise of silver and the gain thereof than fine gold. She is more precious than rubies, and all of the things that you can desire cannot be compared to her. Length of days is in her right hand and in her left hand riches and honor. Her ways are the ways of pleasantness and all her paths are peace. She is a *Tree of Life* to them that lay hold upon her, and happy is everyone that retains her."

King Solomon has been called the wisest king on earth. When he ascended to the Throne, God offered him a choice. He could either have riches and gold, or Wisdom. He chose Wisdom. He explains why he made that choice in Proverbs 4:3-11, "For I was my father's son, tender and only beloved in the sight of my mother. He taught me also, and said unto me, let your heart retain my words, keep my commandments and live. Get Wisdom, get Understanding; forget it not. Do not decline from the words of my mouth. Forsake her not and she shall preserve you; love her and she shall keep you. Wisdom is the principle thing; therefore get Wisdom. And with all your getting, get Understanding. Exalt her and she shall promote you. She shall bring you honor when you embrace her. She shall give to your head an ornament of grace; a crown of glory shall she deliver to you. Hear, oh my son, and receive my sayings; and the years of your life shall be many. I have taught you in the way of Wisdom. I have led you in the right paths."

Wisdom and Understanding are available for us also, if we are willing to search for it. They do not come easily, but they are well worth the search. And the Proverbs of Solomon are the best place to begin that search.

The *virtue* of Binah is *Silence*. Silence and meditation provide the opportunity to tap into the Potentialities of Binah. It is through silence that we begin to learn that Understanding comes from spiritual struggle. Restrictions and limitations dominate our lives, and learning to overcome them provides Understanding and potential Wisdom.

The *vice* of Binah is *Avarice*, which is an insatiable greed for riches and an inordinate desire to hoard wealth and gain. This results from Binah as a reservoir of pure potentiality, which, if out of balance, creates an empty pit, which the greedy person tries to fill.

Realization of Conception in Symbolic Form is active *Intelligence* equilibrated by Wisdom. Wisdom is the Generative Power that is capable of producing Thought. It is not thought, nor is it thinking. It is the power to produce thought. Intelligence is the intellectual capacity to produce thought. It is not thought, nor is it thinking. It is the capacity to produce the thought. When Wisdom and Intelligence, or Understanding, are united, the result is Daath, which is Intellection or Thinking. It is not thought, but it is the action of Thinking. Daath is the source, or outflowing of knowledge, which results from active Intelligence, equilibrated by Wisdom.

Daath

Daath is a point on the pathway between Chockmah and Binah, which is intersected by the pathway between Kether and Tiphareth. Daath is in the Abyss that separates the Supernals from the Lower 7. Daath is the word for *Knowledge*. Daath is not a Sephira because it has no number. Sephira means number, and Sephiroth is the plural or numbers.

In the arena of the Supernals there is only the potential of Thought. Thought is dependent upon its mobility and its transference from one object to another. Thought needs an object or idea outside itself about which to think. Daath is a quasi-emanation point from which those objects and ideas can begin to Emanate in the material plane.

Divine Thought is only pure potentiality. Kether is like a point. Chockmah is like a point extended into a line. A line connects two points but does not define its limits. But it does reveal purpose and Will. A third point is necessary to define the limits of the line. Hence, Binah is the third point, which creates a triangle and a surface. This also creates awareness. However, a fourth point is necessary before manifestation can occur, and that is to provide a 3-dimensional characteristic of depth. It is in Chesed where the 3-dimensional properties of manifestation first become possible.

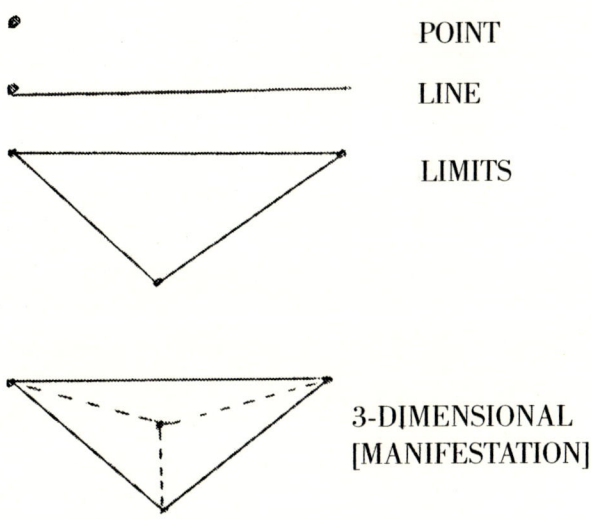

POINT

LINE

LIMITS

3-DIMENSIONAL
[MANIFESTATION]

Manifestation of three dimensional forces

Daath is the process of thinking, which changes Thought from potential to active form. In Chesed, Thought becomes a power and a force. It can construct and build or it can destroy. Thought can be received by the human mind and be used to inspire others, or it can be used to change circumstances and events. It can even be used to control others.

Daath is the secret agent of both generation and regeneration. Daath exemplifies the process of Emanation through the pairing into opposites and their union in a Third. This principle manifests on the material plane by the union of sperm and egg which then becomes a zygote. A zygote is different than either of the two uniting substances from which it came, but is not yet what it shall become, which is an embryo.

In the Archetypal world the concept of Daath is one of realization and illumination. It can transmute learned knowledge into Wisdom and Understanding.

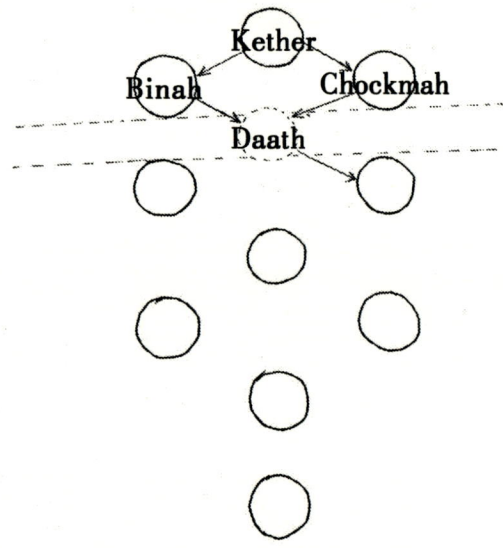

Emanation of thought into the material realm

Another function of Daath is that it provides a "bridge" to the dark side of the Tree of Life, which is the Tree of *Knowledge* of *Good* and *Evil*. The *Light* side of the Tree is referred to as *Able* and the 10 Sephiroth in order from Kether to Malkuth are Consideration, Discretion, Compassion, Love, Industriousness, Care, Dependability, Responsibility, Stability and Obedience. The *Dark* side is known as Cain and the corresponding 10 Sephiroth are Thoughtlessness, Foolishness, Misunderstanding, Hate, Idleness, Carelessness, Frivolity, Indolence, Instability and Anarchy.

Daath is also the bridge between the material and the spiritual planes of existence. The human mind cannot communicate directly with Divine Thought, nor can Divine Thought communicate directly with the human mind. An intermediary is necessary to conduct any communication or contact through the Abyss between the material and the spiritual worlds. This is demonstrated in Romans 8:26-28: **"The spirit also helps our infirmities, for we know not what we should pray for as we ought. But the spirit, itself, makes intercession for us with groanings, which cannot be uttered. And he that searches the hearts knows what is the mind of the spirit because he makes intercession for the saints according to the will of Yahweh."**

Human consciousness cannot transit into the Supernals but must be left behind as Neshamah, our higher Soul, reaches those levels of consciousness. But how can this most easily be done?

In Kether all is One. The Father Principle is revealed in Chockmah, and the Mother Principle is active in Binah. Uniting these two Principles produces a third Principle, the Son. This is the Only-begotten Son, which is The Word or Logos, which was in Union with Abba, the Father in Chockmah. From the Principle of the Son in Daath, material emanation of the Universe was initiated. Our human consciousness is centered mostly in Malkuth. Our Spiritual advancement can most easily be initiated by uniting our consciousness with the Messiah Consciousness of Tiphareth.

This is fully explained in Colossians 1:12-23, 26: "**Giving thanks unto the Father, who has made us meet to be partakers of the inheritance of the saints in light. He has delivered us from the power of darkness and has translated us into the kingdom of his dear Son. In him we have redemption through his blood, even forgiveness of sins. He is the image of the invisible God, the firstborn of every creation. For by him were all things created that are in heaven and that are in earth, visible and invisible, whether they be thrones, or dominions, or principalities, or powers. All things were created by him and for him. He is before all things and by him all things consist. He is the head of the body, the Ecclesia. He is the beginning, the firstborn from the dead, that in all things he might have the pre-eminence. For it pleased the Father that in him should all fullness dwell. And having made peace through the blood of his cross, by him to reconcile all things unto himself . . . whether things in earth or things in heaven. And you that were sometimes alienated and enemies in your mind by wicked works, yet now has he reconciled, in the body of his flesh through death, to present you holy and unblameable and unreproveable in his sight. [That is] If you continue in the faith grounded and settled, and be not moved away from the hope of the good news, which you have heard . . . [This is] the mystery that has been hid from ages and generations but now is made manifest to his saints."**

This mystery has been hidden in the Tree of Life since the beginning of time and has been lost by the confusion of religion today.

Perfection is required to cross the Abyss and this is why testing is so necessary in the advancing spiritual life. Daath is where you are tested before you can advance to the next level of spiritual attainment. Messiah Consciousness at Tiphareth assists you in developing balance in your understanding of right and wrong, or judgment and mercy. Once you have acquired balance your resolve to maintain balance in the face of adversity must be tested. This is when you are led to the level of Daath and the Son becomes the testing block. The Son is responsible for the testing, and also provides the pathway through the testing. This is explained in Hebrews 6:1-2: **"Therefore leaving the principles of the doctrine of Messiah, let us go on to perfection, laying not again the foundation of repentance from dead works, and of faith toward God, of the doctrine of baptisms, and of laying on of hands, and of resurrection of the dead, and of eternal judgment."**

Once the principles of Messiah are understood, it is time to move toward the testing ground of perfection. When perfection is received, enlightenment is perceived and union with the Father is achieved.

It is vital to understand and remember that the Son is always with you during the entire period of testing in Daath. Even during your darkest hour his power is always available for you. Once you have passed the test perfection is yours. That is why James 1:2-4 offers this encouragement, **"My brothers, count it all joy when you fall into diverse temptations, knowing this, that the trying of your faith develops patience. But let patience have her perfect work, that you may be perfect and entire, lacking nothing."**

Once you have passed through the darkest night (Daath) you will become One with the Father, which is achieved through the Son. This is clarified in John 14:23: **"Yahshua answered and said unto him, If a man love me he will keep my words, and my Father will love him and we will come unto him and make our abode with him."**

The Central Pillar of the Tree of Life is the key to spiritual advancement. It provides balance in your life and it provides Union with Divine Thought through the power of Yahshua the Messiah. If you are beginning to understand The Great Work, and you have devoted yourself to pursuing The Great Work, then you are ready to move toward . . .

Completion of the Great Work.

The Supernals

1. The Supernals (Kether, Chockmah, and Binah) are the first 3 Sephiroth of the Tree of Life.
2. The Supernals are separated from the lower 7 by a gulf called the Abyss.
3. Divine Thought is the eternal pattern of the Universe emanated from the Ain Soph or the Endless World. The pattern constitutes the form of a man and is called Adam Kadmon.
4. *Kether*, the Corona or Crown, is Divine Will and is the head of Adam Kadmon.
5. The name of God in Kether is EHEIEH (I AM THAT I AM) The Eternal Existent One.
6. The *virtue* of Kether is *Completion of the Great Work.* There is no vice in Kether.
7. *Chockmah*, or Wisdom, is Divine Intellectual Power, or the ability to generate intellect.
8. Chockmah is the beginning of duality, and is Abba, The Father.
9. Chockmah is also the level of The Logos or The Word as the energizing Principle.
10. The name of God in Chockmah is YAHOWEH (Jehovah) or YAH, The Eternal.
11. The *virtue* of Chockmah is *devotion.* Again, there is no vice
12. *Binah,* or Understanding, is Divine Intellectual Capacity.
13. Binah is Marah the Mother or Cosmic Womb, and is a reservoir of Intellect.
14. The name of God in Binah is YAHOWEH ELOHIM, the Creator through the Masculine/Feminine Principle.
15. The *virtue* of Binah is *silence.* The *vice* is *avarice* or *greed.*

16. *Daath,* which means knowledge, is a point in the Abyss midway between Chockmah and Binah.
17. Daath is the point of union between Chockmah and Binah, which produces the *process* of thinking.
18. Daath is a bridge to the dark side of the Tree of Life.
19. Daath is also a bridge between the material and the spiritual planes of existence.

Chapter 12

ADAM KADMON

Adam Kadmon is the idea or the intellectual aggregate of the whole Universe contained unevolved in Divine Thought. Divine Thought conceived of the idea of creating a family of sons and daughters with whom to share knowledge and experience. The children of Divine Thought would learn Understanding and Wisdom through the experience of sharing with one another and with Divine Thought. Experience requires a body, and the body of Divine Thought would become the Manifested Universe.

Divine Thought is One. The whole Universe is One, which is developing itself into the many. The Unity of Divine Thought became the Duality of Elohim or male and female. From the Duality came the many sons and daughters as sparks of life. These sparks of life manifest in the material Universe as *souls* that come to gain *experience* through sharing and exercising the qualities of Justice and Mercy with others. Once the soul has gained sufficient Understanding and Wisdom, it can return to the Unity of Divine Thought while simultaneously retaining its *purified identity!*

I John 3:2-3 says, **"Now are we the sons (and daughters) of**

Yahweh, and it does not yet appear what we shall be. But we know that when he shall appear we shall be like him for we shall see him as he is. And everyone that has this hope in him purifies himself, even as he is pure."

Adam Kadmon as the body of Divine Thought is the macrocosm while Man is the microcosm. This spiritual Ideal is the beginning of all of creation whereas material man is the end or completion of that creation. Superior Adam supplies everything while Inferior Adam receives all. Superior Man descends from the undivided down to the lowest level of power, while Inferior Man ascends from the lowest depths to the higher realms. Adam Kadmon descends and flows downward into his own nature so as to gain experience, then returns back to the Unity which he possesses within himself. And this is repeated again and again on an individual basis with each Soul of Man.

This Kabbalistic Principle is also revealed in I Corinthians 15:45-49, **"And so it is written, the first man, Adam was made a living soul, the last Adam a quickening spirit. Howbeit that was not first which is spiritual, but that which is natural, and afterward that which is spiritual. The first man is of the earth** *earthy;* **the second man is the Master from Heaven. As is the earthy such are they also that are earthy, and as is the heavenly such are they also that are heavenly. And as we have born the image of the earthy we shall also bear the image of the heavenly."**

The first Adam was created in the image of Adam Kadmon, as a living soul contained in a body that is subject to death. The second Adam descended directly from the Adam Kadmon, and possessed within himself the power of Immortality, which is the "image of the heavenly." It is possible for *us* to acquire the image of the heavenly as well.

The Logos, which the Bible calls *The Word*, dwells in Divine Thought and is the world of Ideas. The Word is the Chief of Intelligence, the Primitive man, the Adam Kadmon. The Word is the Creator and presently occupies the place of the Supreme

Being. In the Supernal regions The Supreme Being and the Savior are interchangeable, as they are One at that level. That is why Paul could write in a letter to Titus, in Titus 1:2-4, 2:13, **"In hope of eternal life which God, that cannot lie, promised before the world began. But has in due times manifested his Word through preaching, which is committed unto me according to the commandment of God our Savior. To Titus, my own son after the common faith; grace, mercy and peace from God the Father and the Master Yahshua Messiah our Savior . . . Looking for that blessed hope and the glorious appearing of the great God and our Savior Yahshua Messiah."**

The Word is Yahshua the Messiah, and he has a most unusual mission. The human family was created in the image of Adam Kadmon, but it presently falls way short of being in that image today. The Word was manifested in Yahshua for the purpose of restoring the human family back into the Image of Adam Kadmon. And this restoration project is being accomplished on a one to one basis, one soul at a time.

Colossians 1:26-27 explains how this is being accomplished: **"[We preach] the *mystery* which has been *hid* from ages and from generations, but now is being made manifest to his saints, to whom Yahweh would make known what is the riches of the glory of this mystery among the gentiles, which is *Messiah in you* . . . "**

One reason the Bible is so hard to understand is that it is written in such long sentences. But the bottom line of this understanding is that there is a power that can be added to your life that begins to regenerate your mortal body into one that can become Immortal. That power is the Messiah (The Word) within you.

This is put another way in John1:12, **"But as many as received him [the Word] to them gave he power to become the sons of Yahweh, to them that believe on his name."**

The power to become the children of Yahweh flows from the

Supernals and centers in Tiphareth. When you allow that power to balance your life, it is Yahshua that provides that balance. Remember that Yahshua, as the Word, is the Chief of Intelligence, and the *source* of Wisdom and Understanding, Judgment and Mercy, and Balance. Those are very powerful energies that can *transform* your life.

It is put like this in John 14:6, 11-14, **"Yahshua said unto him I am the way, the truth, and the life; no one comes unto the Father but by me . . . Believe me that I am in the Father, and the Father in me; or else believe me for the very Work's sake. Truly, I say unto you, he that believes on me the works that I do he shall do also, and greater works than these shall he do, because I go unto my Father. And whatsoever you shall ask in my name, that will I do, that the Father may be glorified in the Son. If you shall ask anything in my name I will do it."**

Yahshua lived a normal human life much as we do today. He got hungry, he got tired and when he stubbed his toe it hurt just like ours. The difference is that he had an open channel to the Father from the very beginning. He suffered anger, frustration, disappointment and even betrayal. And he had to deal with it just like we do. He experienced all of these and much more so that he would know what it was like to be human.

When we accept the energizing force of Yahshua to live within us we can draw upon his experience to help us with our own. And we can also draw upon the power of his will, and his love, and his sense of justice and mercy. And in this way we become One with Messiah as he is One with the Father. And that makes us One with the Father as well. From that point forward, we are on the road to perfection. Suffering, and how you handle it, leads to perfection when Yahshua is in you.

The Bible sums up the whole message of Dynamic Kabbalah in Colossians 1:15-19 **"[The Son] is the image of** the invisible God, the firstborn of every creature. For by him were all things created, that are in heaven, and that are in earth, visible and invisible, whether they be thrones, or dominions, or principalities, or powers. All things were created by him and for him. And he is before all things and by him all things consist. And he is the head of the body, the Ecclesia, who is the beginning, the firstborn from the dead, that in all things he might have the pre-eminence. For it pleased the Father that in him should all fullness dwell."

The Great Work of Yahshua the Messiah is quite simple. As Messiah, and the Word, he is the Adam Kadmon, which is the image of the invisible God. As Chief of Intelligence he is the head of Adam Kadmon. As Yahshua, the man, he is the spiritual Adam or the perfected and restored Man. He is also the head of the *Ecclasia*, a body of people *"called out"* from among the masses for a special purpose. The *Ecclesia* [not church] is composed of sons and daughters of Yahweh who have Yahshua dwelling in them.

The unit of humanity is made in the image of Adam Kadmon, as are the individuals who make up humanity. But humanity has fallen from its original estate. The unit of *Ecclesia* is made in the *restored* and *regenerated* image of Adam Kadmon with Yahshua as its head. The individuals of Ecclesia have Immortality of body and soul within their grasp.

Why is Yahshua important? Because from Tiphareth he has access through the Abyss to the Father in the Supernal region. And Yahshua within you can access the Father from your own Tiphareth (heart) on your own internal Tree of Life.

Humanity has fallen so deeply into Malkuth that confusion, unrest, uncertainty, guilt, resentment, and victimization blind most individuals. It is time for individuals to take responsibility for their own lives and well being. If you believe that Yahshua has the power to change your life then now is the time to begin that inner dialog with him. He can and will remove all of the Kharma that has accrued to your account since the beginning of the Universe until today. Not only from this life, but also from all of the lives in your past. Yahshua is the Bread of Life who came to remove your Bread of Shame. You can begin with a clean record today. Yahshua explained how in John 6:28,29,32-35,51, **"Then they asked him, what shall we do that we might do the works of God? Yahshua answered and said unto them, This is the work of God that you believe on him whom he has sent . . . my Father gives you the true bread from heaven. For the bread of God is he who came down from heaven and gave life unto the world. They asked him, Master, forever give us this bread. Yahshua said unto them, I am the bread of life; he that comes to me shall never hunger, and he that believes on me shall never thirst . . . I am the living bread which came down from heaven; if anyone eat of this bread he shall live forever; and the bread that I give is my flesh which I will give for the life of the world."**

Yahshua later explained that he was using a metaphor to illustrate spiritual life. Everyone knew that bread, or food, is what energizes the body and keeps it alive. The Soul or the spiritual life also needs food or sustenance to keep it energized. He is illustrating that he is that energizing force in the Universe, and is the source of Immortality or eternal life. Sacrificing his body would release the power of the Logos into the Universe for all who would accept it. This regenerating power is available to you, if you choose to accept it. Be bold! Step up to the plate! Seize the moment! You will be amazed at how much better you will feel when you realize that you have an inner partner to share your burdens, sorrows, hurts and joys.

Do it now!

Adam Kadmon

1. Adam Kadmon is the idea, or the ideal pattern for the Universe and everything in it.
2. The purpose of the Universe is to create sons and daughters of Divine Thought.
3. There is a material Adam subject to death, and a spiritual Adam possessing Immortality made in the image of Adam Kadmon.
4. The Word, or the Logos is Adam Kadmon.
5. Yahshua the Messiah is the head of Adam Kadmon and of the Ecclesia (aggregate of sons and daughters of Divine Thought).
6. The Ecclesia is the restored and regenerated Adam Kadmon.
7. Yahshua can gain access to the Father through the abyss for us from Tiphareth, and thereby add us to the Ecclesia.
8. It is time to assume responsibility for your own life.
9. Let Yahshua begin to transform your life from within!

Bibliography

Allen, James. *As A Man Thinketh.* Barnes and Noble, 1992.

Andrews, Ted. *Simplified magic.* St. Paul, Llewellyn Publishers. 1991.

Berg, Dr. Phillip S. *Kabbalah for the Layman, Vol. I, II, and III.* New York, Research Center of Kabbalah. 1991.

Berg, Dr. Phillip S. *Power of Aleph Beth.* New York, Research Center of Kabbalah. 1988.

Chopra, Deepak. *The Seven Spiritual Laws of Success.* San Raphael, Ca. Amber-Allen Publishing. 1993.

Covey, Steven R. *The 7 Habits of Highly Effective People.* New York. Fireside Publishers. 1990.

Davies, Paul. *Superforce.* New York, Simon and Schuster. 1984.

Dyer, Dr. Wayne W. *The Secrets of Manifesting Your Destiny.* Niles, Illinois, Nightingale-Conant. 1994. (Cassette tape series based on book of same name.)

Franck, Adolphe. *The Kabbalah.* New York, Bell Publishing. 1940. (Translation from French.)

Gawain, Shakti. *Creative Visualization.* Mill Valley, Ca., Whatever Publishers, Inc. 1985.

Gonsalez-Whippler, Migene. *A Kabbalah for the Modern World.* New York, Bantam Books. 1977.

Hay, Louise L. *You Can Heal Your Life.* Carson, Ca., Hay House. 1990.

Kaku, Michio and Jennifer Thompson. *Beyond Einstein.* New

York, Anchor Books Doubleday. 1995.

Moyers, Bill. *Healing and The Mind.* New York, Doubleday Publishers. 1993.

Myss, Carolyn, Ph.D. *Anatomy of the Spirit.* New York, Crown Publisher. 1996.

Pearsall, Paul, Ph.D. *The Heart's Code.* New York, Broadway Books. 1988.

Pert, Candice, Ph.D. *Molecules of Emotion.* Carson, Ca., Hay House. 1997.

Pike, Albert. *Morals and Dogma.* Richmond, Va., L.H.Jenkins, Inc. 1950.

Regardie, Israel. *A Garden of Pomegranates.* St. Paul, Llewellyn Publishers. 1992.

Richo, David, Ph.D. *Unexpected Miracles.* New York, Crossroad Publications. 1998.

Sitchin, Zecharia. *The 12ᵗʰ Planet.* New York, Stein and Day Publishers. 1976.

Strong, James. *Strong's Exhaustive Concordance of the Bible.* New York, Abingdon Press. 1975.

The Bible. [Various translations used.]

Walker, Barbara, G. *The Woman's Encyclopedia of Myths and Secrets.* San Francisco, Harper and Row Publishers. 1983.

Weinberg, Steven. *The First Three Minutes.* New York, Bantam Books. 1984.

Printed in the United States
1504200001B/303